Membranes and Circadian Rythms

Springer
Berlin
Heidelberg
New York
Barcelona
Budapest
Hong Kong
London
Milan
Paris
Santa Clara
Singapur
Tokyo

Th. Vanden Driessche (Editor)
J.-L. Guisset, G.M. Petiau-de Vries (Co-Editors)

Membranes and Circadian Rythms

With 28 Figures and 6 Tables

Springer

PROF. DR. THERESE VANDEN DRIESSCHE
Université Libre de Bruxelles
Département de Biologie Moleculaire
46, avenue des Ortolans
1170 Bruxelles, Belgium

DR. JEAN-LUC GUISSET
Université Libre de Bruxelles
Pool de Physique
Faculté de Sciences, Campus Plaine C.P. 222
50, avenue F.D. Roosevelt
1050 Bruxelles, Belgium

DR. GHISLAINE M. PETIAU-DE VRIES
Université Libre de Bruxelles
Chimie Generale I, Faculté de Médecine
Campus Solbosh C.P. 160/18
50, avenue F.D. Roosevelt
1050 Bruxelles, Belgium

ISBN-13:978-3-540-60101-2 e-ISBN-13:978-3-642-79903-7
DOI: 10.1007/978-3-642-79903-7

Library of Congress Cataloging-in-Publication Data. Membranes and Circadian Rythms / Th. Vanden Driessche (ed.), J.-L. Guisset, G.M. Petiau-de Vries (co-eds.). p. cm. Includes bibliographical references.ISBN-13:978-3-540-60101-2(alk.paper)I.Cell Biology. II.Driessche, Th.Vanden. QQR387.P38 1996 616.9'25075-dc20 DNLM/DDLC for Library of Congress 95-9578 CIP

SPIN: 10507630 27/3136 – Printed on acid free paper

Preface

Rambling of an elderly biochemist

Most biochemists of my generation , who were trying to discover the pathways of metabolism, simply ignored membranes; or regarded them as a nuisance. Think of the difficulties experienced in studies on cytochromoxidase which one could not separate from « insoluble material » or again of the desperate efforts during a quarter of a century to unravel oxidative phosphorylation without paying much attention to lipidic membranes, altough the system was known to be associated with them. Hence the amazement and the general skepticism that met at first Mitchell's theory, which was giving membranes the central function they deserve in oxidative phosphorylation and photosynthesis. This, I believe, was a turning point; enzymologists thereafter became aware of the importance of the membranes.

Neurophysiologists, of course, had long been interested in the outer cell membrane with its electrical properties and the ion potentials. Histologists and electronmicroscopists also, who observed inside the cell.organelles of which membranes are essential components: nucleus, nucleoli, mitchondria, lysosomes,Golgi apparatus, endoplasmic reticulum, chloroplasts. For them at least, a cell did not look like a mere bag full of enzymes and small molecules; they knew, they could see that it is a highly structured system divided into many compartments by membranous formations.

Thanks to newer developments of Molecular Biology and to the progressive resolution of macromolecular structures, we know, now, that a membrane is much more than a bilayer of phospholipids; it is a complex structure, an organized machinery with ion pumps, energy coupling systems, selective gates, photoreceptors, allosteric hormone-receptors. These control the exchanges of metabolites and of signals between the outer medium and the inside of the cell, or between inner organelles and the cytosol. Membranes are not static structures; many of them are involved in a complex traffic, an endless reorganization during secretion, phagocytosis, pinocytosis, cell division, cell movements, cell contacts and fusion, and all kinds of adaptative processes; they also carry recognition molecules, carbohydrates or otherwise, responsible for cell adhesion or controlling cell proliferation. Membranes are indeed involved in every physiological process of the cell.

The numerous inner cell organelles are as many compartments, as many little worlds with their own characteristic enzyme assortments and metabolic activities;

exchanges between these physically separate systems depend strictly on the properties of the membranes, regulated by subtle cybernetic interactions. As a result, the kinetics of the integrated cell metabolism is of a formidable complexity, with membranes intervening everywhere.

In studies of global kinetic phenomena like biological rhythms, membranes are crying for more consideration than they have been given so far.

One must therefore be grateful to the Editor of this book for presenting together up to date reviews by specialists of different aspects of circadian rhythms in different organisms, for putting special emphasis on membranes and on the fascinating idea that they may coordinate different systems submitted to circadian rhythms.

Macromolecules or pigments embedded in the plasma membrane are indeed constrained by the membrane structure; their properties, their state of aggregation, their association into higher order complexes are modulated by the state of the membrane. Circadian changes in plasma membrane structure could therefore affect together the activities of various molecules involved in signal transduction thus modifying in a concerted, integrating manner the activities of otherwise independent systems.

It will not be easy to untangle all the biochemical and physical interactions involved in circadian rhythms, but good working hypotheses are taking shape, and powerful methods of genetics and molecular biology are coming to the rescue. From a conceptual point of view, valuable contributions might come from theoretical studies on rhythmic processes in complex systems, or on the logics of stable oscillatory behaviour versus damping, steady state, or chaos; they might point to the most significant parameters and help to orient- or to select among- more or less intuitive hypotheses proposed by experimentalists.

Out of the enormous amount of experimental data already assembled, some will eventually prove irrelevant, and among the proposed interpretations some may be ill-fated, but clear conclusions are bound to emerge. In any new field of research, especially with as complex a system as a living organism, explorers are at first groping around in the confusing accumulation of observations and experimental data, until here and there some unexpected result or some fortunate rapprochement of independent observations triggers an illuminating idea which solves part of the puzzle, or opens the way to new perspectives.

The present book, in bringing together data from different sources and in putting the emphasis on membranes and their probable coordinating function, is bound to play an important part in the maturation of problems related to circadian rhythms and in their eventual solution.

Hubert Chantrenne

Contents

Contributors

Asard Han Department of Biology, University of Antwerp (RUCA) Groenenborglaan, 171, B-2020 Antwerpen, Belgium

Bonzon M. Laboratoire de Physiologie végétale, Université de Genève, 3, Place de l'Université, CH-1211 Genève 4 , Switserland

Brody Stuart Department of Biology, 0116, University of California, San Diego, 9500 Gilman Drive, La Jolla, CA 92093-0116, USA

Bruguerolle Bernard Laboratoire de Pharmacologie Médicale et Clinique, Université d'Aix-Marseille II, Faculté de Médecine. 27,rue Jean Moulin, F-13385 Marseille CEDEX 5 France

Caubergs Roland Department of Biology, University of Antwerp (RUCA) Groenenborglaan, 171, B-2020 Antwerpen, Belgium

Coté Gary G. Department of Molecular and Cell Biology, U-125, University of Connecticut, Storrs, CT 06269-3125 USA

Edmunds Leland N., Jr Division of Biological Sciences. University of New York at Stony Brook , Stony Brook, New York 11794-5200, USA

Gaspar Thomas Laboratoire d'Hormonologie Fondamentale et Appliquée. Institut de Botanique. Université de Liège, Bat. B-22, Boulevard du Rectorat, 27, B-4000 Liège, Belgium

Goldbeter Albert Unité de Chronobiologie Théorique, Faculté des Sciences, Université Libre de Bruxelles, Campus Plaine C.P. 231 ,Bd du Triomphe, B-1050 Bruxelles, Belgium

Greppin H. Laboratoire de Physiologie Végétale, Université de Genève, 3, Place de l'Université, CH-1211 Genève 4 , Switserland

Guisset Jean-Luc Pool de Physique, Faculté des Sciences, Université Libre de Bruxelles, Campus Plaine C.P. 222 ,Bd du Triomphe, B-1050 Bruxelles, Belgium

Homblé Fabrice Laboratoire de Physiologie Végétale Faculté des Sciences, Université Libre de Bruxelles, Campus Plaine C.P.206/2 , Bd du Triomphe , B-1050 Bruxelles, Belgium

Kiefer S. Laboratoire de Physiologie Végétale, Université de Genève, 3, Place de l'Université, CH-1211 Genève 4 , Switserland

Lakin-Thomas Patricia Department of Plant Sciences, University of Cambridge, Downing street, Cambridge CB2 3EA, United Kingdom

Normann J. Biologisches Institut II, Universität Freiburg, Schänzlestr. 1, D-79104 Freiburg, Germany

Penel C. Laboratoire de Physiologie Végétale, Université de Genève, 3, Place de l'Université, CH-1211 Genève 4 , Switserland

Petiau-de Vries Ghislaine M. Chimie Générale I, Faculté de Médecine, Université Libre de Bruxelles, Campus Solbosh C.P. 160/18, 50, av.F.D. Roosevelt, B-1050 Bruxelles, Belgium

Ruiz-Fernandez S. Biologisches Institut II, Universität Freiburg, Schänzlestr. 1, D-79104 Freiburg, Germany

Vanden Driessche Thérèse Département de Biologie Moléculaire. Université Libre de Bruxelles. 46, avenue des Ortolans, B-1170 Bruxelles, Belgium

Wagner E. Biologisches Institut II, Universität Freiburg, Schänzlestr. 1, D-79104 Freiburg, Germany

1 Introduction and Overview

T. Vanden Driessche
Département de Biologie moléculaire, Université Libre de Bruxelles, 46, av. des Ortolans, 1170 Bruxelles, Belgium.

It was our aim, in undertaking this book, to emphasize the importance of circadian rhythms (CR) in modulating plasma membrane functions and consequently, in many cases at least, in regulating signals perception.

Various aspects of membranes and CR have been addressed in different organisms. The selection of the organisms by the authors is generally based on their amenability to particular types of research.

Several authors concluded their chapters with stimulating hypotheses involving membranes in the operation of CR or the mechanism of rhythmicity itself.

Membrane lipids are studied in chapter 2 in relation with CR. The fatty acid composition of Neurospora membranes in wild type and mutants is abundantly documented. Neurospora is an Ascomycota which displays a basically stable period in its conidiation rhythm, contrary to other fungi displaying periods much longer than 24 h, often greater than 100 h (in Aspergillus for example). Neurospora is possibly at a later evolutionary stage. It displays a CR in linoleic and linolenic acid contents. Coté, Lakin-Thomas and Brody documented lipid mutants and examined in parallel the conidiation CR expression. They conducted these studies in the wild type, in a number of mutants and under a variety of light (essentially blue light) and temperature conditions. They took into consideration the different lipid types of the plasma membrane as well as the lipids of the mitochondria. They made clear that defects in lipid metabolism always correlate with major CR alterations.

The problem of photoperception and nature of the photoreceptors is discussed: the plasma membrane and the endoplasmic reticulum being both candidates for up to now not definitively identified photoreceptors (see also chapter 7).

The possibility that changes in lipid composition could explain temperature compensation is an attractive hypothesis. The authors carried out experiments aimed at manipulating membrane fluidity. This property of CR is important and its mechanism not understood. Some fatty acids lengthened the period in Neurospora. Mitochondria also are subjected to changes in lipid composition.

Interestingly, several mutations of the mitochondrial genome reverse this period lengthening, showing that mitochondrial metabolism may affect Neurospora CR.

Eventually, however, the authors had to conclude that no simple relationship could be found between membrane composition and period length.

I would like to suggest that the results, rather than being interpreted in the general frame of a single oscillator, with input and output pathways, could be reexamined in the hypothesis that more than one oscillator operate, one at least, intimately connected with the plasma membrane and another one with mitochondria.

After all chapters were in shape for printing (except the overview), a paper of Rosbach and collaborators (Huang et al., 1995) appeared with new results and interpretation of temperature compensation in Drosophila. It is based on Hardin et al. (1990) initial model for CR generation, model described by Goldbeter in chapter 4. Accordingly, both per mRNA and PER protein amounts are cycling in the cytoplasm and involve a negative feed back loop onto the nuclear located gene. Consequently, it is the rate of dimers formation which determines the circadian cycle period: as a transcriptional regulator, the protein PER contains a so-called PAS domain comprising a basic helix-loop-helix, a dimerization motif. The per^L mutation lies in this region. The authors introduced a further change in one amino acid of the protein (PER^LR). Dimerization in vivo and in vitro of these various proteins at different temperatures and behaviour of the flies showed differences. The authors observed that in addition to the interactions between PAS domains of PER proteins, there are intramolecular interactions within the protein, between the PAS domain and sequences of the COOH-terminal domain. Huang and colleagues propose that, since both types of interactions, inter- and intramolecular, bind to the PAS domain, they are in competition. Both interactions should change similarly with temperature. Since dimerization favours PER activity, which is opposed by the intramolecular interaction, and since the PER^L mutation might give rise to different temperature coefficients, a change would result in the balance between the reactions and the components accompanied by a loss in temperature compensation. The authors indeed observed a reduced dimerization degree in per^L mutants. Neurospora and Drosophila display a similar rhythmic behaviour and analogous mutations were obtained in both species. In addition, Aronson et al. (1994) observed simultaneous alterations in period length and temperature compensation in a Neurospora frequency mutant affected in a single open reading frame. It would be very interesting to submit the temperature compensation model to experimental verification in Neurospora while, at the same time, following possible parallel modifications in fatty acid composition. In any case, the hypothesis cannot be of universal relevance: it is meaningless for those CR which do not require transcription (see chap 9 for example).

Glycosylation (Petiau-de Vries, chapter 3) provides proteins with additional properties or specificities. Glycosylation could intervene in CR operation, at least in some cases. The frequent glycosylation of receptors suggests a function to be explored.

Among the many provocative results given by Petiau-de Vries is the growth hormone receptor, whose total de-glycosylation suppresses binding activity.

Receptors are systematically reviewed, as well as hormones, when glycosylation appears to be involved. It is remarkable that a primary step (although not the following ones) in both rhodopsin, the higher animals visual pigment of the photoreceptor, and phytochrome, a major plant photoreceptor, acts through a G protein: there is apparently, a conservation of an early transduction step (see review by Chory, 1994).

Various glycosylated polymers of great diversity make up the extracellular matrix of both plant and animal cells. According to Engel et al. (1994), the extracellular matrix proteins are organized in different domains which could assemble in various ways, adding to the diversity. Common assembly domains and similarity in sequence and structure suggest common ancestry.

Petiau-de Vries gives several examples related to cancer or other illnesses showing that glycosylation is potentially important in pharmacology, and may introduce a temporal dimension to the picture. Erythropoietin, for example, is a haematopoietic growth hormone displaying isoforms differing in their oligosaccharide moiety. The hormone's level in serum is itself periodically changing, following a CR whose parameters appear to change in some cases of illness. As suggested by the author, a way of manipulating these molecules might be to modify the phase or amplitude of a possible CR.

Also in the plant world, glycosylation is very important. In the unicellular alga, Acetabularia, it has been shown that glycosylation metabolism is modulated by circadian rhythmicity.

The intriguing proteoglycan coded by the per gene in Drosophila was not discussed: no clear role could finally be defined (see Baylies et al., 1993) and a possible function questioned by Takahashi (1992). These large molecules, which become external to the plasmalemma, might interfere with cell communication. Proteoglycans are probably synthetized according to a CR, if the whole and very long per gene is transcribed as a block.

One of the major subjects of research nowadays -of very intensive research along different lines- is the cell cycle control. The trigger is the membranar adenylate cyclase which initiates a complex cascade of events. Other molecules may reach the membrane, unduly activate the cascade and lead, in some cases, to pathological conditions. Basically, two questions have been asked. First, what makes the cell division cycle (CDC) oscillate periodically in many cases (Goldbeter, chap. 4)? And second, how does the circadian clock gates the CDC, at least in numerous species and tissues (Edmunds, chap. 5)?

Goldbeter (chapter 4) discusses four biochemical oscillations. He first summarizes the main features of one of the simplest and best understood biological rhythm, glycolysis, a high frequency rhythm. He then explains the cyclic AMP oscillations, also a high frequency rhythm, inducing aggregation in the slime mold Dictyostelium. He underlines the detailed mechanism of membranar receptor desentization, a process critical for the oscillation. He turns then to the CDC, which may be ultradian rather than circadian in many Eukaryotes. Finally, last but not least, the circadian rhythm generating mechanism, based on Drosophila molecular genetics, in particular experimentation on the per gene and its PER protein product. The cycling per mRNA and PER protein as a basis for the fly's circadian rhythmicity were initially proposed by Hardin and colleagues (1990), but further experiments have been carried out by several authors. Goldbeter integrates the results in a new, more complete model. Concerning this model, however (as well as that of Hardin et al.) I would like to underline that, although Drosophila and possibly other Eukaryotes data fit in very well, the proposed CR generating mechanism cannot be applied to Acetabularia which displays undamped circadian rhythms whatever nucleate or anucleate (see chap. 9). In other words, it is not universal in nature. All four rhythms modelized by Goldbeter appear to be of the limit cycle type. All involve at least one feed-back loop and, except glycolysis, they partly depend on phosphorylation/dephosphorylation processes.

Edmunds (chap. 5) presents a synthesis of the extensive research carried out in his Lab on the CDC and CR, a research based on Euglena as model system.

Euglena was considered to be a Protist or an algal flagellate. However, the recently revised classification by Margulis (1993) brought the genus in the kingdom of Protoctista.

Edmunds aimed at understanding the nature of the circadian clock and the mechanism which couples CR and CDC.

Edmunds, on the one hand demonstrates that Euglena cell cycle is expressed periodically as a typical CR, which can free run, be entrained and phase shifted by synchronizing agents. On the other hand, the CDC influences the CR: for example, when perturbed, it dramatically reduces the amplitude of the biphasic CR in cyclic AMP (cAMP). The author hypothetized that cAMP -which controls certain limiting steps of the CDC- might also play a coupling role. Extensive experimentation, making use of an array of specific inhibitors, brought the notion that cAMP acts downstream from the CDC.

Edmunds examined whether cGMP plays a role in the mediation of the circadian rhythm action on the CDC. His experimental findings suggest that cGPM could take part in the control of the cAMP CR.

At the heart of the problem is the nature of the links between the CR and the CDC. Considering the times at which cAMP affects the CDC, the author hypothetized that cAMP could regulate the cell cycle progression at two crucial transition control points: in the G1 phase and at the G2/M boundary.

The results were incorporated in a model for the gating of the CDC by the circadian clock and the model argumented. In addition, new data were presented, strengthening Edmund's view. They concern changes in the physical state of cyclin B and $p34^{cdc2}$ at different circadian times.

If we consider Euglena and Acetabularia cell cycles it is clear that they do not compare since the duration of their CDC is respectively one or a few days and one year). However, it is interesting, to note that very young Acetabularia have a fairly high cyclic AMP content, an extremely low one during the rapid growth phase (period of time during which cAMP oscillates) and then an increasing one again up to very high values (Minder and Vanden Driessche, 1978). Moreover, if the presence of cAMP is demonstrated in plants, the actual role of cAMP is still unclear (Brown and Newton, 1981). Perhaps will Edmund's results bring researchers to pay a greater attention to its possible influence on mitosis.

In chapter 6, Homblé gives a synthetic overview of the major ion transport processes in plant plasma membranes, including brief and interesting comments on techniques. The H^+-ATPase and channels are the major membrane carriers. The former belongs to the classical P class ATPases which are phosphorylated during their reaction cycle whereas the fast channels exist in one or more states and may be regulated biophysically (voltage gated) or biochemically (ligand gated). The plasma membrane H^+-ATPase gene has been cloned. For Saccharomyces cerevisiae, a DNA sequence analysis predicts a K^+ channel homologous to the animal ones.

Various plant movements for which ion movements are of critical importance are examined: the nyctinastic movement of leaves or leaflets achieving periodically "day" and "night" positions as well as the rapid movements displayed by some species. Nyctinastic movements are due to turgor changes of the two halves of pulvini for which K^+ fluxes are mainly responsible. In addition, there is a complex distribution of the latter (Schrempf and Mayer, 1980) and they are influenced by a number of regulators (phytohormones for example). Our personal view is that these movements are extremely complex, due to the specific morphology, number of channels and receptors and make up of the cytoskeleton and cytoplasmic proteins (Vanden Driessche, 1980).

Spontaneous oscillations in action potential have been found to occur in a number of plants. In Nitella for example, the actual sequence is described. The spontaneous repetitive action potential and its interesting theory are described.

The opening-closing oscillations of stomata movement are examined in relation with H^+, K^+, Cl^- and also Ca^{2+} and anions. The movement is influenced by a variety of regulators, light for example (as indicated in chap. 7), which triggers the activity of the H^+-ATPase.

The oscillations in membrane electric potential imply involvement of membrane transport processes. Oscillations of 1 h period can be generated by H^+-ATPases and oscillations of 15 min by K^+ channels. From these considerations, Homblé discusses two possible rhythm generation models. First, coupling involving feed-back loops between photosynthetic reactions and membrane potential difference may generate oscillations. Second, the particularly attractive model of Gradman and coworkers in which the oscillation arises from coupling between carriers and voltage- dependent channels. In this case, the plasma membrane could be an autonomous oscillator. However, these oscillations are either high frequency oscillations or oscillations of very short period.

To complete the picture, Homblé discusses spatial patterning in plants -an important parameter in morphogenesis. Patterning is correlated with membrane potential differences and coupling between H^+-ATPase and ion channels.

Light reception in higher plants -photosynthesis excluded- is dealt with by Asard and Caubergs (chap. 7). It is often believed that there are two major photoreceptors: the phytochrome (whose localization is a matter of debate) and the plasma membrane located cryptochrome, which now appears to be only one of several blue light receptors.

Phytochrome is involved in numerous physiological responses. The authors discuss the recent results concerning its nature, mode of action and essentially cytoplasmic localization. The authors emphasize, however, that the involvement of G proteins in phytochrome transduction might give an answer to the question of phytochrome localization and its relation with membrane by transient attachment to G proteins.

They discuss blue light perception and recall the most important blue light morphogenetic responses and their critical role in circadian rhythmicity synchronization.

Several blue light receptors are now considered to exist in plants: the membranar complex flavin-cytochrome b (the so-called cryptochrome), possibly but not definitively demonstrated to be associated with the light-dependent absorbance change (LIAC), other plasma membrane putative receptor molecules, carotenoids, oxidoreductases. Flavins may be part of the oxidoreductase. There also exist as free flavins (or pterins in their photoexcited form).

Several blue light receptors may operate in conjunction: some processes are dependent on two cooperating photoreceptors. Moreover the cooperation of blue and red/far red activation is required in a number of cases.

Many physiological blue light responses, unambiguously involving the plasma membrane are reviewed. They include circadian turgor changes in pulvini responsible for nyctinastic movements and stomatal movement. Both movements have a circadian component. Both imply ion fluxes which are dealt with in great detail by Homblé (chap.6).

An early step in blue light transduction is the phosphorylation of a specific 120 kDa protein. The molecules possibly involved in the process are considered.

Free flavins, which, however, may bind to a protein not yet defined, might be implicated in circadian rhythm synchronization (they were principally studied in fungi, see chap.2)

The general conclusion is that the field is moving fast but that there are still many questions to be answered.

After completion of the chapters editing work (except for this overview), two papers from Kay and colleagues appeared (Millar et al. 1995, a and b), solving the open problem of which from blue or red light synchronize CR in higher plants. They imagined a clever way of visualizing CR in Arabidopsis: they hooked the circadian regulatory sequence of the cab gene (chlorophyll a/b binding protein) to the luciferase gene of firefly. The former is responsible for the photosynthesis CR of numerous higher plant species. The enzyme makes luciferin to glow. Suffice then to spray the plants with luciferin to observe its glow when the luciferase gene is activated by the circadian promotor, at certain phases of the 24 h cycle. The authors recorded cotyledons glowing times, in response to the cab gene activity and determined the CR period. The rhythm is in synchrony with cotyledons movements. The method allows easy detection of plants out of phase, i.e. CR mutants. In the wild type, the period is lenghtened in DD (30-36 h) whereas under constant light, either red or blue, the period is much shorter, of about 25 h.

The authors then introduced the construction, cab gene-luciferase gene into Arabidopsis phytochrome-deficient mutants. The resulting plants display in constant red light a long period (although not as long as that of control in DD). In contrast, in blue light, they display a period very similar to that of the wild type. This and other experiments demonstrate that, in Arabidopsis, both red and blue light exert a control on CR synchronization.

B. Bruguerolle's chapter (Chap.8) bears directly on clinical research, taking into account his studies on membranes and their variations in the 24 h cycle. Indeed, since the pioneer work of the sixties, the notion of chronopharmacology proved to be extremely rewarding (see recent books: Reinberg, 1994 covering basic concepts and Touitou and Haus, editors, 1992, reviewing major chronopharmacological domains).

The CR in the physiological effect of a drug is integrative: it implicates knowledge of the drug mode of action, its pharmacokinetics and its desired and toxic effects, all varying with the time of day.

In the first place, Bruguerolle details the various steps between drug administration and its target, emphazising the multiplicity of intervening CR, specially the membrane CR. He reports several studies obtained by his group concerning plasma protein binding (total plasma proteins themselves varying along the 24 h scale) and chronokinetics of several anaesthetics in animals and humans. Some differences were evidenced between healthy and diseased people and subjects of different age.

The author demonstrated that the changes observed in some anaesthetics effects can be traced to both plasma protein binding and to erythrocyte permeability.

A number of drugs display circadian rhythmicity in penetrating into erythrocytes.These are considered as models for the study of temporal changes in membrane permeability.

Bruguerolle reports temporal changes in tissue binding of several anaesthetics in various tissues.

He underlines differences occuring between drugs that experimental work of his group has evidenced and discusses the possible consequences in clinical application.

Chapter 9 (Th. Vanden Driessche) is devoted to extraordinary membranes, the thylakoids or photosynthetic membranes of the chloroplasts of three Protoctista, Acetabularia, Euglena and Gonyaulax. Thylakoids are the membranes of closed saccules, internal to the chloroplast. They comprise the light-harvesting pigment-protein complexes, the molecules involved in the photosynthetic light reactions and the photochemical reaction centers. Thylakoids are subjected to periodic ultrastructure conformation change which result in a high amplitude photosynthesis CR. The rhythm may involve a periodically translated, undetermined, protein. No every day transcription is required. The case illustrates wonderfully well the great importance membranes can have.

Chapter 10 is specifically devoted to the effects of synchronizing environmental factors on CR and the significance of membranes. For Wagner and collaborators, in the course of evolution, organism not only adapted to a cyclic energy supply (on a daily and on an annual basis) but corresponding metabolic activities became the timer for circadian rhythms and for developmental processes.

They present arguments suggest that the plasma membrane functions are rhythmic, including phytochrome pelletability. "Rhythmic stimulation of plasma membrane-bound receptors could feed back on cellular energy metabolism which in turn could rephase sensitivity of plasma membrane-bound receptors and metabolic functions".

Indeed, they evidenced a CR in Mn^{2+}-dependent binding of alkaline peroxidases to microsomes (vesicles formed from the plasma membrane), time-dependent binding of phytochrome to plasma membrane, and rhythmic sensitivity of the receptors which, according to their experiments, depends on the cell's energy status.

They compared three Chenopodium ecotypes for which they defined common parameters in the critical photoperiod for flowering. The results imply a signal transmitted from leaves to apex, possibly of electrical or electrochemical nature. It results eventually in frequency-coded inducing signals. The apical cells display a 20 h rhythm in cells division and a 30 h one in leaf originating floral stimulus. The former might limit the action of the latter. Their interaction could result in a CR in stem elongation rate. Moreover, a change in the leaf CR due to the photoperiod reaching a critical day, would modify the rhythms relation which might produce morphogenetic changes.

In the last chapter, Vanden Driessche and collaborators concentrate on the plasma membrane of Acetabularia, belonging to the Protoctista. It turns out to be an illustration, in a particular case, of notions brought about by the preceeding chapters: basic rhythmic activity of the plasma membrane: transmembrane bioelectric potential, fatty acid components, glycosylation degree; to these, by comparison with other systems we should add phosphorylation of some proteins within the membrane or in proximal cytoplasm. The plasma membrane contains the photoreceptors and several other receptors, to phytohormones, to insulin , as well as the first step(s) of their transducing pathways. Sensitivity to light and phytohormones varies periodically, as well as sensitivity of some steps of these pathways to drugs specifically targeted against particular components. Redox related processes are dependent on the plasma membrane and some components are subjected to circadian rhythmicity. This simple enumeration points to the plasma membrane as a signal integrating organelle. This has also been the view of several authors (Greppin and collaborators, 1991, Homblé, this volume, among others). Consequently, the various pathways of a transducing network are activated by several receptors and signals. They should bring about concerted action in transduction and harmonious distribution of energy. Membranes are also influenced by the cytoplasm , its energy status, its pH, the concentration of some molecules and ions, Ca^{2+} in particular. Thus the membrane is functioning in taking into account both external conditions and internal status.

Moreover, we propose that circadian rhythmicity has a further regulatory function: it modulates the plasma membrane activity and imposes, in doing so, suppression of one degree of freedom (see below, Rhythms and models). In other words, CR exert a stringent regulation of the plasma membrane functions.

Perspectives

Not all plasma membrane functions have been investigated so far in taking CR into account. My guess is that in a near future several chapters will have to be added. Some chapter should concern the various intracellular organelles (only chloroplasts thylakoids have been studied): the endoplasmic reticulum, the Golgi apparatus might be subjected to CR. This would of course bear on intracellular traffic, exo- and endocytosis etc... The nuclear membrane should also be examined from this view.of point. Of particular importance will be the chapter on the cytoskeleton, the knowledge of which is rapidly expanding. The plasma membrane provides both anchoring and signals reception, relayed in several cases by the cytoskeleton. In this context, it should be recalled that a protein related to human beta-spectrin has been identified by Michaud et al. (1991) in plant cells. Of course, if a membrane does vary with time in the course of the 24 h, the possibility exists of a circadian modulation of specific functions (for the influence of membranes, see Bennett and Gilligan, 1993; Luna and Hitt 1992).

It should be underlined that there are many cases where membrane attachment is required for activity, such as the now classic mechanism of protein kinase C activation by diacylglycerol (Bell, 1986). A recent example is the one in which the ligand activated tyrosine receptor requires membrane attachment (or clustering) for activity (Davis et al., 1994). If indeed a CR modulates the plasma membrane, there is a probability that circadian changes occur in receptor activity. This type of process possibly explains changes in sensitivity to hormones or chemical effects in some cases. Plasma membrane conformation changes might affect, among many others, integrins. They are instrumental in interlocking on the plasmamembrane. They contain themselves proteoglycans. Fibronectins, which bind to the integrin receptor are, together with laminin, extracellular matrix proteins. They may transfer chemical signals to the cell provided that the integrin receptor also exerts a mechanical force at the cell surface (see Nowak, 1994).

Of great interest in the context of this book is the demonstration by Eskin and his collaborators of membranar proteins modulated by circadian rhythmicity and possibly implicated in the oscillatory mechanism. The Aplysia (a marine mollusk) isolated eyes displays a CR dependent on both transcription and translation (Koumenis and Eskin, 1992). In an effort to determine whether translation is part of a feed-back loop in the "central" oscillator, the CR generating mechanism or whether if its influence is indirect, they isolated proteins whose synthesis or phosphorylation is altered in opposite ways by light and by serotonin (5-HT), both entraining the CR. A 40 kDa protein was specially promising. Sequence similarity, immunochemistry and immunohistochemistry showed that the protein is an annexin. Annexins are Ca^{2+}-dependent phospholipoprotein, also called lipocortins. This annexin appears to be involved in the chain of events leading to expression of the CR of optic nerve impulses. Some annexins of the synexin type may act as Ca^{2+} channels; some may function by inhibiting phospholipase A2 (PLA2) activity. Indeed, Calignano et al. (cited by Takahashi et al., 1993) obtained evidence that the 40 kDa protein extracted from Aplysia modulates PLA2 activity. This could be of significant importance since PLA2 regulates arachidonic acid metabolism. Additional effects are not excluded (Takahashi et al. 1993). Annexins are not restricted to animals: they have been identified in plants (Blackbourn et al. 1992).

Rhythms and models

Our ignorance concerning the CR generating mechanism and CR relations to major aspects of cells and rhythms is reflected by the different views expressed by several authors.

Coté and colleagues (chap. 2) discuss the data obtained concerning their hypothesis on CR temperature compensation (to which I added and discussed a just out of press model).

Goldbeter's devotes his chapter (4) to experimentally based models, including a CR generating one based on transcription/translation and a feed-back loop.

Edmunds (chap.5) concentrates on a model coupling CR to the CDC.

Homblé (chap. 6) underlines the oscillation generating power of ionic movements through the plasma membrane, describing two models, one of which is a membranar autonomous oscillator.

According to Wagner et al.(chap.10), the key for CR generation is the following cycle: the energy metabolism influences the energy-transducing biomembrane state. Its receptors, in turn control the cellular energy metabolism and re-initiate the cycle.

Furthermore, the authors suggest that photoperiodic induction relies upon the interaction of two rhythms located respectively at the apex and in the leaves, from which the floral stimulus is assumed to be generated (possibly through electrical coupling).

There are some common points between several models, including mine (Vanden Driessche, 1994). My view, however, is somewhat at variance. Coherent with our chap. 11, Vanden Driessche and collaborators, is that interactions between two networks result in overt circadian rhythms. The first network concerns oscillations in receptors containing membranes (photoreceptors and several receptor types) and their transducing pathways on the one hand and a multiplicity of coupled oscillators (metabolic and others) on the other. Among the metabolic oscillators are glycolysis as well as cycles of phosphorylation/dephosphorylation, glycosylation/deglycosylation etc... An autonomous oscillation of the membrane is not excluded. The oscillating cycles need not be the same throughout development. Plasma membrane, coupling, the receptors allows synchronization. Moreover, CR exert a control function in gating membrane various activities in time.

Future Directions

Finally a new chapter in Chronobiology should be open: chronopharmacology applied to agriculture.

In plant cultures, Koukkari (University of Minnesota, St Paul) met some success (personal communication), but this was not apparently the case in Europe, specially in Belgium. There are a few major ways in which herbicides are effective (as already explained by Corbett, 1974 in his pioneer work): (1) those interfering with photosynthesis and (2) those interfering with plant growth. These are prominently interacting with phytohormone physiological roles. We know that both types of action are subjected to circadian rhythmicity (Vanden Driessche, 1983 as concerns photosynthesis and work of my Lab, reflected in the periodic sensitivity of Acetabularia towards three major hormones (Chap. 11).

As concerns mammals, chronobiological studies should be conducted on the general model used for humans in order to evidence times of greater sensitivity or of maximum deleterious effects when pesticides are used. As for insects, specific sensitivities and phase manipulation should also be studied, as they are in Hayes' Laboratory in Beltsville, Maryland for defense against insects. These considerations are relevant for environment.

A very large number of pesticides and herbicides are free radicals generators, either immediately or after a delay of some days or weeks, having been either metabolized or sujected to modifications in mammals.

References

Aronson BD,Johnson KA,Dunlap JC(1994)Circadian clock locus frequency/ protein encoded by a single open reading frame defines period length and temperature compensation. Proc Natl Acad Sci USA 91:7683-7687

Baylies MK,Weiner L,Vosshall LB,Saez L,Young MW(1993)Genetic, molecular and cellular studies of the per locus and its products in Drosophila melanogaster. In: Young MW (ed.) Molecular genetics of biological rhythms. Marcel Dekker New York Basel Hong Kong :123-153

Bennett V,Gilligan DM(1993)The spectrin-based membrane skeleton and micron-scale organization of the plasmamembrane. Ann Rev Cell Biol 9:27-66

Blackbourn HD,Barker PJ,Huskisson NS,Battey NH(1992)Properties and partial sequence of plant annexins. J Plant Physiol 99: 864-871.

Brown EG,Newton RP(1981)Cyclic AMP and higher plants. Phytochemistry 20:24-63

Chory J(1994)Phytochrome signal transduction. Current Biol 4: 844-846.

Corbett JR(1974)The biochemical mode of action of pesticides. Acad Press London New York pp.330

Davis S,Gale NW,Aldrich TH,Maisonpierre PC,Lhotak V,Pawson T,Goldfarb M,Yancopoulos GD(1994)Ligands for EPH-related receptor tyrosine kinases require membrane attachment or clustering for activity. Science 266:816-819.

Engel J,Efimov VP,Mauer P(1994)Domain organization of extracellular matrix proteins and their evolution. Development Supplement 35-42.

Greppin H,Bonzon M,Crespi P,Crèvecoeur M,Degli Agosti R,Penel C, Tacchini P(1991)Communication in plants. In:Penel C,Greppin H (eds)Plant signalling, plasma membrane, and change of state Imprimeria Nationale, Genève.

Hardin PE,Hall JC,Rosbach M(1990)Feedback of the Drosophila period gene product on circadian cycling of its mRNA levels. Nature 343:536-540

Huang ZJ,Curtin KD,Rosbach M(1995)PER protein interactions and temperature compensation of a circadian clock in Drosophila. Science 267:1169-1172

Koumenis C,Eskin A(1992)The hunt for mechanisms of circadian timing in the eye of Aplysia. Chronobiol Internat 9:201-221

Luna EJ,Hitt AL(1992)Cytoskeleton-plasma membrane interactions. Science 258:955-964

Margulis L(1993)Symbiosis in cell evolution. Microbial communities in the archean and proterozoic eons. WH Freeman and Co.New York

Michaud D,Guillet G,Rogers PA,Charest PM(1991)Identification of a 220 kDa membrane-associated plant protein immunologically related to human beta-spectrin. FEBS Letters 294:77-90
Millar AJ,Carré IA,Strayer CA,Chua N-H,Kay SA(1995 b)Circadian clock mutants in Arabidopsis identified by luciferase imaging. Science 267:1161-1163.
Millar AJ,Straume M,Chory J,Chua N-H,Kay SA(1995 a)The regulation of circadian period by phototransduction pathways in Arabidopsis. Science 267:1163-1166
Minder C,Vanden Driessche T(1978)Changes in the cyclic AMP content during growth and development of Acetabularia. Differentiation 10:165-170
Nowak R(1994)Matrix work wins acclaim (Meeting brief). Science 263:30-31
Reinberg A(1994) Les Rythmes Biologique Mode d'emploi.Flammarion pp 170.
Schrempf M,Mayer WE(1980)Electron microprobe analysis of circadian changes in K and Cl distribution in the laminar pulvinus.Zeit Pflanzenphysiol 100:247-255
Takahashi JS(1992)Circadian clock genes are ticking. Science 258:238-240
Takahashi JS,Kornhauser JM,Koumenis C,Eskin A(1993)Ann Rev Physiol 55:729-753
Touitou Y,Haus E(eds)(1992)Biological Rhythms in Clinical and Laboratory Medecine. Springer Verlag Berlin Heidelberg New York London Paris Tokyo Hong Kong Barcelona Budapest
Vanden Driessche T(1980)The seismonastic leaf movement of Mimosa pudica and the nyctinastic leaf movement. In:De Greef (ed.) Photoreceptors and Plant Development. Proc Ann Europ Symp Photomorphogenesis, Antwerpen 1979:599-616
Vanden Driessche T(1984)Photosynthesis, circadian rhythms and herbicides. In: Sybesma C.(ed) Photosynthesis 1983, IV.3:301-304
Vanden Driessche T(1994)Circadian rhythms in three unicellular organisms. The peculiarities of the organisms, the evidence brought on rhythms and their specific practical problems. Biol Rhythm Research 25:345-386

2 Membrane Lipids and Circadian Rhythms in *Neurospora crassa*

Gary G. Coté[1], Patricia L. Lakin-Thomas[2], and Stuart Brody[3].
[1]Department of Molecular and Cell Biology, U-125, University of Connecticut, Storrs, CT 06269-3125 USA
[2]Department of Plant Sciences, University of Cambridge, Downing Street, Cambridge CB2 3EA, United Kingdom
[3]Department of Biology, 0116, University of Calfornia, San Diego, 9500 Gilman Drive, La Jolla, CA 92093-0116 USA

Summary

The possibility that membranes are involved in circadian rhythmicity in the ascomycete fungus *Neurospora crassa* is discussed. *Neurospora* is an ideal organism for testing this possibility; knowledge of its biochemistry and genetics is substantial, and mutants affecting circadian rhythms and/or lipid biochemistry are available.

The *Neurospora* circadian oscillator receives input on temperature and illumination from the environment. Evidence suggests that the *Neurospora* blue-light photoreceptor may be membrane-localized. Studies using *Neurospora* inositol and choline auxotrophs, however, appear to rule out hydrolysis of phospholipids by a phospholipase C or a phospholipase D in transduction of light

---

Abbreviations: DAG, *sn*-1,2-diacylglycerol; Ins(1,4,5)*P*3, *myo*-inositol 1,4,5-trisphosphate; PtdCho, phosphatidylcholine; PtdEtn, phosphatidylethanolamine; PtdIns, phosphatidylinositol; PtdIns(4,5)P_2, phosphatidylinositol 4,5-bisphosphate; PtdOH, phosphatidic acid. Fatty acids are represented in the text by x:y notation in which x indicates the total number of carbons in the acyl chain and y indicates the nimber of double bonds. Positions of the double bonds are indicated by the standard device of using superscripts to a Greek letter delta preceding the x:y notation. The delta prefix is ommitted for the major unsaturated fatty acids of *Neurospora*: Δ^{9}-18:1 (oleic acid), $\Delta^{9,12}$-18:2 (linoleic acid) and $\Delta^{9,12,15}$-18:3 (α-linolenic acid). The major saturated fatty acids of *Neurospora* are 16:0 (palmitic acid) and 18:0 (stearic acid).

signal information from the photoreceptor to the oscillator. The mechanism of temperature signal transduction remains unknown, but temperature-induced changes in membrane properties could play a role.

Involvement of membrane lipids in the mechanism of the oscillator itself is suggested by correlation of lengthened period with abnormal membrane composition in the *Neurospora* mutants *cel*, *prd-1* and *chol-1*. The *cel* mutant is blocked in fatty acid synthesis; both the period of its rhythm and the fatty acid composition of its membranes are sensitive to environmental conditions, no simple relationship between its membrane composition and its period length has yet been described. The *chol-1* mutant is also altered in lipid synthesis; it is a choline auxotroph unable to synthesize phosphatidylcholine unless choline is supplied. The *prd-1* mutant was isolated on the basis of long circadian period; it has an abnormal membrane fatty acid composition, although its primary biochemical lesion is unknown.

The levels of linoleic and linolenic acid in the membranes of *Neurospora* vary with a circadian rhythm, and similar rhythms in fatty acid composition have been reported in a number of species of plants and animals. These rhythms could be part of the circadian oscillator mechanism, or part of an ouput pathway from the circadian oscillator.

Temperature compensation of membrane fluidity by enzymatic alteration of membrane lipid composition may play a role in the temperature compensation of rhythmicity, and this possibility is discussed. Temperature compensation of rhythmicity appears to be lost in three different mutants of *Neurospora*: *cel*, *chol-1*, and null alleles of *frq*. The *cel* and *chol-1* mutants are defective in lipid synthesis, and are likely to show abnormal adjustment of membrane composition with temperature. The *frq* locus was originally identified on the basis of altered period length. There is no evidence for defective lipid metabolism in *frq* mutants, but such a defect has not been ruled out.

2.1 Introduction

The possibility of a connection between circadian rhythms and membranes has been under discussion in the literature for many years. A seminal paper, by Njus et al. (1974), proposed a membrane model for the oscillator incorporating the well-known phenomenon of temperature compensation of membrane fluidity to account for the temperature compensation of circadian period. Membrane models of circadian oscillators have been reviewed by Engelmann and Schrempf (1980) and by Edmunds (1988). The evidence for a role for membranes in the *Neurospora crassa* clock has been reviewed recently (Lakin-Thomas et al. 1990). In brief, this evidence comes from studies on the effects of "membrane-active" substances, such as ionophores and D_2O, on the phase and/or period of the rhythm, as well as studies reviewed here on the relationship between fatty acids, membrane composition and the clock. This review will not attempt to cover the topic of membranes in general, but will focus on membrane lipids in particular and their potential roles in circadian rhythmicity in *Neurospora*. As will be discussed here, the evidence supports a strong connection between membrane lipids and temperature compensation of the period in *Neurospora*.

The *Neurospora* circadian system has been reviewed recently (Lakin-Thomas et al. 1990) and a review of the molecular genetics of this system can be found in Dunlap (1993). *Neurospora* is a filamentous fungus that displays a circadian rhythm of conidiation (asexual spore formation) when growing across the surface of solid agar medium. It can also be grown in liquid medium, and although it does not display a rhythm of conidiation under these conditions, it is possible to transfer pieces of the mycelium to solid agar medium and demonstrate from the phase of the subsequent conidiation rhythm that the circadian oscillator is functional in liquid medium. *Neurospora* is easily grown in defined medium and its lipids are easily labeled with radioactive tracers and extracted for analysis. *Neurospora* has been a model experimental organism in genetics laboratories for many years as well, and a large collection of nutritional, morphological, and chronobiological mutants is available. The combination of biochemical and genetic methods developed for *Neurospora*, and the detailed information already available on this organism, makes *Neurospora* particularly suited to the analysis of the role(s) of membrane lipids in circadian rhythmicity.

Table 1. Properties of the *Neurospora* circadian rhythm in wild type and selected mutants defective in rhythmicity and/or lipid metabolism

Mutation	Period, hr (22°C)	Q_{10}	Phase resetting by light (22°C)	Phase resetting by temperature
Wild-type	21	1	Normal	Normal
frq[1]	16	1	Increased response	N.D.
frq[7]	29	1.3	Decreased response	Decreased response
frq[9] and *frq*[10]	15-30	2	N.D.	N.D.
prd-1	26	1	Decreased response	N.D.
inl (25 μM inositol)	21	N.D.	Increased response	Normal
inl (250 μM inositol)	21	N.D.	Normal	Normal
chol-1 (0.5 μM choline)	52	2.5	N.D.	N.D.
chol-1 (100 μM choline)	21	1	N.D.	N.D.
cel (minimal)	21	2.2 below 22°C	Normal	N.D.
cel (40 μg/ml 16:0)	21	4.4 below 18°C	N.D.	N.D.
cel (40 μg/ml 18:2)	40	2.3 below 26°C	Decreased response	N.D.

See text for description of mutants. Growth supplements are indicated in parentheses following mutant designations; N.D., not determined.

The lipid composition of membranes is an important determinant of membrane structure and function, and membranes might be involved in clock functions in several ways: 1) Clock components that are not membrane-bound may require the physical structure or compartmentation provided by membranes for proper functioning; membranes may thus play supporting roles in the clock. 2) Membrane composition can influence the activity of integral membrane proteins, and clock components may be membrane-bound. 3) Membranes themselves might be clock components, and the structure or function of membranes may oscillate. Specific membrane lipids might also play a direct role as active participants in the circadian clock. They might be components of signaling pathways providing input to regulate the oscillator or providing output from the oscillator, or they might participate in the mechanism of the oscillator itself.

This review is divided into four sections, covering input, oscillator, output, and temperature compensation. The first section will ask the question: Could phase resetting of the oscillator by changes in temperature or light pulses act through effects on membrane lipids? The second section will describe mutations that change the properties of the oscillator and cause changes in lipid metabolism as well. The third section will look at circadian rhythms in lipid composition as an output of the oscillator. In the final section, the possible connection between temperature compensation of the period and lipid composition will be discussed. The data to be presented in this review are derived largely from biochemical analyses of cultures under various environmental conditions, and analyses of mutants with known biochemical lesions. These studies demonstrate the strengths of the *Neurospora* system for a biochemical and genetic approach to understanding circadian rhythmicity. Table 1 summarizes much of the genetic evidence to be presented in this review.

2.2 Lipids and Input

2.2.1 Phase Resetting by Temperature

In nature, organisms monitor temperature and light signals and respond by adjusting the phase of their circadian rhythms to maintain synchrony between these rhythms and the cycle of day and night. In the laboratory, the response of rhythms to environmental input is monitored as changes in phase resulting from environmental manipulation. Phase resetting in *Neurospora* by temperature manipulation has been reviewed (Lakin-Thomas et al. 1990). Both temporary changes (temperature pulses) and permanent changes (temperature steps) are effective, as are both temperature increases and decreases. A recent study (Gooch et al. 1994) confirms these basic observations and clearly shows that phase shifting can be either strong (type 0) or weak (type 1), depending on the size of the temperature change.

Abrupt changes in temperature affect the fluidity of biological membranes, increasing fluidity if the temperature increases or decreasing fluidity if the temperature decreases. The effect is temporary, since cells remodel their membranes, adjusting lipid composition to restore normal fluidity. If membrane lipids or membrane proteins are components of the clock this temporary fluidity change could alter the normal cycling of the clock, producing phase shifts. It has been suggested (Njus et al. 1974) that normal maintenance of membrane fluidity might underlie temperature compensation (see discussion below). Phase shifts might then result from temporary failure of temperature compensation following abrupt changes in temperature and membrane fluidity. Alternatively, membrane lipids and proteins might not be clock components, but a membrane-localized temperature sensor might detect changes in membrane fluidity and transduce the information to clock components proper. It is also possible, of course, that temperature effects on rhythmicity are not mediated by changes in membrane properties.

At present, there is neither experimental evidence to link changes in membrane fluidity with phase resetting induced by temperature changes, nor any evidence against such a linkage. One way to test such a linkage would be to test for correlations between defects in membrane remodeling following a temperature change and defects in subsequent phase resetting. Clock mutants could be examined for defects in phase resetting induced by temperature, and then for defects in membrane remodeling. Only the clock mutants frq^1 and frq^7 (*frequency*), with shortened and lengthened periods, respectively, have been examined for their phase-resetting responses to temperature (Nakashima 1987; Rensing et al. 1987). Although the results differed between laboratories, the long-period mutant frq^7 appears to be less responsive than frq^+ to both cold pulses and heat pulses (Nakashima 1987; discussed by Lakin-Thomas et al. 1991). Mutants with defects in lipid synthesis or degradation could also be examined for defective membrane remodeling and then tested for altered phase resetting in response to temperature changes. The *cel* and *chol-1* mutants discussed below are good candidates for such studies.

2.2.2 Phase Resetting by Light

Introduction. The second important environmental variable with input to the clock is light: the phase of the *Neurospora* clock can be reset by pulses of blue light, and continuous light damps the rhythm. The blue-light photoreceptor is likely to be a flavin linked to a *b*-type cytochrome, but this photoreceptor has not been positively identified. Nor has any signal transduction pathway been identified as the link between the photoreceptor and the oscillator (see Lakin-Thomas et al. 1990 for a review of blue-light signal transduction in *Neurospora*). Membrane lipids may affect the blue-light input pathway in two ways: indirectly, by forming the environment surrounding membrane-bound components of the pathway, or directly, as active players in the signaling mechanism.

The Blue-Light Photoreceptor. Borgeson and Bowman (1985) attempted to identify the subcellular location of the putative flavin-cytochrome photoreceptor complex by assaying blue light-induced absorbance changes in various membrane fractions. Absorbance spectra characteristic of *b*-type cytochromes were found in the plasma membrane, endoplasmic reticulum, and mitochondria. The plasma membrane cytochrome most resembled the cytochrome linked to the photoreceptor, based on its sensitivity to inhibitors and similarities between its absorbance spectrum and the action spectrum for blue-light phase resetting. This work indicates that the blue-light photoreceptor complex may be located in the plasma membrane, although a minor cytochrome in the endoplasmic reticulum or mitochondria cannot be ruled out. It should be noted that the work of Klemm and Ninnemann (1978) raises questions about the use of light-induced absorbance changes to identify the photoreceptor for rhythmicity: they found no correlation in starved mycelia between absorbance changes and phase resetting.

Borgeson and Bowman (1990) have also assayed the photoreduction of cytochromes in membrane fractions from *frq* mutants. Significant differences were found in the cytochromes of the endoplasmic reticulum: in frq^9, a null mutant, this fraction showed no blue-light reduction of the cytochrome, and in frq^1 the endoplasmic reticulum cytochrome showed an altered spectrum (relative to the wild-type) after blue-light treatment. These authors suggest a connection between the altered temperature compensation of the *frq* mutants and the altered *in vitro* responses to blue light: The endoplasmic reticulum contains a lipid desaturase complex that includes a *b*-type cytochrome, and if this desaturase is defective in the *frq* mutants, it could account for altered period and altered temperature compensation (through altered lipid composition) as well as for altered blue-light responses. If this hypothesis is correct, it might be predicted that the *frq* mutants would have altered fatty acid composition, but this is apparently not the case (Feldman et al. 1986; Coté and Brody 1987a). Although this hypothesis does not require that the endoplasmic reticulum cytochrome be an *in vivo* blue-light photoreceptor, Lyudnikova et al. 1989), noting a correlation between photoinduction of carotenoids and fatty acid desaturation during temperature acclimation, proposed that the flavoprotein of the endoplasmic reticulum desaturase complex might function as the blue-light photoreceptor for carotenoid induction in *Neurospora*.

The Blue-light Signal Transduction Pathway: Inositol Lipids. Beyond the photoreceptor, the blue-light signaling pathway in *Neurospora* has not been identified. The lipid-based signaling mechanisms described in animal cells are attractive candidates for transduction of blue-light signals from the photoreceptor to the oscillator. In the best-studied such pathway, the inositol lipid signaling system (Berridge 1987), activation of a receptor results in hydrolysis of the lipid PtdIns(4,5)P_2 and release of two intracellular messengers, the neutral lipid DAG which activates protein kinase C, and the water-soluble headgroup Ins(1,4,5)P_3 which mobilizes intracellular calcium. The existence of the inositol-requiring *inl* mutant in *Neurospora* has allowed a critical test of the hypothesis that phase resetting by blue light is mediated by PtdIns(4,5)P_2 hydrolysis. Inositol depletion of an *inl* strain growing on solid agar medium was found to change the threshold dose of blue light required for strong phase resetting; however, inositol depletion unexpectedly increased, rather than decreased, sensitivity to light, by several orders of magnitude (Lakin-Thomas 1992). A change in sensitivity to a resetting stimulus could occur by a change in the input pathway or by a change in the amplitude of the oscillator (Lakin-Thomas et al. 1991). Resetting by temperature pulses was not affected by inositol depletion, suggesting that there was no change in the amplitude of the oscillator. Inositol depletion therefore seemed to alter the sensitivity of the blue-light input pathway, but not in the direction predicted if an inositol lipid signaling system were directly involved.

The levels of inositol metabolites in the *inl* strain were assayed by labeling with [^{3}H]inositol during growth (see Table 2). On solid agar medium, inositol depletion increased the level of the putative signaling precursor PtdIns(4,5)P_2 by about 70%; however, in liquid medium, more severe inositol depletion decreased the level of PtdIns(4,5)P_2 to 25% of control values (Lakin-Thomas 1993a,b). Light sensitivity was measured in liquid medium by assaying the threshold dose of blue light required to elicit strong phase resetting in mycelial disks transferred to solid agar medium in race tubes, but no change in light sensitivity was found in response to inositol depletion. These results indicate that there is no direct correlation between the levels of the putative signaling molecule PtdInsP_2 and light sensitivity. When the levels of Ins(1,4,5)P_3 were assayed during exposure to doses of blue light that were effective in shifting the phase, no significant changes were seen in Ins(1,4,5)P_3 levels from 5 sec to 2 min of light exposure. This result has been essentially confirmed by Prior et al. (1994), who found no effect of a 60 sec white light pulse on the levels of inositol phosphates or inositol lipids in *Neurospora*. Taken together, these results make it unlikely that an inositol lipid signaling system acts as the blue-light input pathway in *Neurospora*.

There is as yet no explanation for the increase in light sensitivity with inositol depletion on solid medium. Inositol depletion may have secondary effects on processes required for light perception, such as the synthesis of pigments associated with the photoreceptor. Alternatively, it may affect light sensitivity by

Table 2. Inositol metabolites and light sensitivity

Solid medium:	250 µM inositol	25 µM inositol
Inositol metabolites (pmol/mg dry weight):		
PtdIns	4420	3790
PtdIns(4,5)P_2	5.0	8.6
Ins(1,4,5)P_3	7.1	6.1
Phase-resetting threshold (µmol/m^2):		
	25,000-50,000	25

Liquid medium:	200 µM inositol	25 µM inositol
Inositol metabolites (pmol/mg dry weight):		
PtdIns	2500	240
PtdIns(4,5)P_2	24	6.9
Ins(1,4,5)P_3	0.6	0.3
Phase-resetting threshold (µmol/m^2):		
	36-70	36-70

affecting lipids other than inositol phospholipids. The effects of inositol depletion in *Neurospora* include an increase in the level of phosphatidylserine, an increase in lipid peroxidation, increased microviscosity of mitochondria, and changes in inositol-containing sphingolipids (see Lakin-Thomas 1992 for references). In animal cells, inositol depletion results in an accumulation of DAG (Berridge et al. 1989). Any of these effects might alter the activity of membrane-associated signaling systems.

The Blue-Light Signal Transduction Pathway: Other Candidates. In addition to the inositol phospholipid signaling system, there are other well-documented lipid signaling systems in animal cells based on the hydrolysis of PtdCho and other lipids (Billah and Anthes 1990). External stimuli may activate phospholipase C or D, releasing DAG or PtdOH, respectively. The PtdOH produced by phospholipase D may either be rapidly converted to DAG or may act as a second messenger itself (Billah and Anthes 1990). Phospholipase A_2 may also be activated, releasing arachidonic acid and lysophosphatidylcholine (Liscovitch 1992). The activation of PtdCho hydrolysis is usually assayed in animal cells by monitoring the increase in DAG and/or choline or phosphorylcholine (depending on the phospholipase activated). The choline-requiring *chol-1* mutant of *Neurospora* was labeled during growth with [^{3}H]choline, and the levels of choline metabolites were assayed after 2, 5 and 60 s of blue light. No systematic changes were seen in the levels of labeled PtdCho or free choline, and no phosphorylcholine or lysophosphatidylcholine were found (P.L. Lakin-Thomas, unpublished results). When similar cultures were grown with [^{3}H]16:0 to label the fatty acid components of lipids, no change was seen in the level of labeled DAG in response to blue light (P.L. Lakin-Thomas, unpublished results). It seems that none of the well-known lipid signaling systems found in animal cells are activated by blue light in *Neurospora*.

Protein kinases, calcium or other ion fluxes, and cyclic nucleotides have also been proposed as blue-light-induced intracellular messengers. Rensing et al. (1993) have reported light-induced changes in protein phosphorylation and in cyclic nucleotide levels. Oda and Hasunuma (1994) have assayed phosphorylation in *Neurospora* microsomal preparations, and have found rapid blue light-stimulated phosphorylation of several proteins of 15 kDa. Whichever signaling system turns out to be the blue-light input pathway to the oscillator, it may be membrane-associated and therefore its activity may be influenced by the lipid composition of the membrane in which it sits.

Effects of Clock Mutants on Phase Resetting. As described above, the clock mutants *frq*^{1} and *frq*^{7} affect phase resetting by temperature pulses. Phase resetting of the *frq* mutants by light pulses was investigated by Dharmanada (1980), and the data were re-analyzed by Lakin-Thomas et al. (1991). The short-period *frq*^{1} and *frq*^{2} mutations increased the response to light, while the long-period *frq*^{3} and *frq*^{7} mutations decreased the response. The *frq*^{7} mutation also decreased the response to phase resetting by pulses of the protein synthesis inhibitor cycloheximide (Dunlap and Feldman 1988). There is therefore a correlation between period and phase resetting in the series of *frq* mutants such that an increase in period is associated with decreased response to phase resetting by several stimuli. We have proposed (Lakin-Thomas et al. 1991) that the *frq* mutations may change the amplitude of the oscillator and that this single effect can account for both the change in period and the change in resetting responses. Dharmananda (1980) also assayed phase resetting by light in other clock mutants, and found a decreased response in the long-period *prd-1* (*period*) mutant. The *cel* mutation, described below, may also fit this pattern under conditions where the period is long, the resetting response to light pulses is decreased. These mutations may therefore have no effect on input pathways but rather may affect the oscillator mechanism. This is in contrast to the *inl* mutation described above, which affects phase resetting by light but has no effect on the period or on resetting by temperature pulses under the same conditions and may therefore affect the light input pathway.

2.3 Lipid Metabolism and the Oscillator

2.3.1 The *cel* Mutant

Introduction. If clock components are intimately associated with membrane lipids we might predict that mutants unable to properly adjust membrane composition so as to preserve membrane fluidity might show defects in circadian rhythmicity. The *cel* mutant of *Neurospora* is a good candidate for such a mutant. Defects in rhythmicity and lipid metabolism in this strain which are reviewed here are summarized in Tables 1 and 3.

The *cel* (fatty acid chain elongation) mutant requires saturated fatty acids for optimal growth (Henry and Keith 1971); however, the requirement is not absolute, as it grows, albeit slowly, without supplement (Elovson 1975). The fatty acid synthase of *cel* has levels of bound 4'-phosphopantetheine cofactor which are less than 2% of wild-type levels (Elovson 1975), and it thus has greatly

Table 3. Alterations in rhythmicity and lipid metabolism in the *cel* mutant

Alterations in rhythmicity
• Temperature compensation lost[a]
• Period and temperature compensation sensitive to supplemental fatty acids[b]
Alterations in lipid metabolism
• Fatty acid synthesis impaired[c]
• Incorporation and oxidation of supplemental fatty acids increased[d]
• Bulk phospholipid fatty acid composition altered[e]
• Bulk phospholipid fatty acid composition sensitive to supplemental fatty acids[e]
• Mitochondrial phospholipid fatty acid composition altered[e]

[a]Mattern et al. 1982.
[b]Reviewed in Lakin-Thomas et al. 1990.
[c]Elovson 1975.
[d]Coté 1986.
[e]Coté and Brody 1987b.

reduced rates of fatty acid synthesis (see Fig. 1a). Of several proteins incorporating pantothenic acid only the fatty acid synthase had deficient incorporation in the mutant (Lakin-Thomas and Brody 1985b). This suggests that the primary biochemical defect in *cel* may be either the fatty acid synthase itself or a cofactor-attaching enzyme specific for the fatty acid synthase. Circadian rhythmicity in the *cel* mutant is profoundly disturbed.
Temperature compensation is lost at temperatures below 22°C as described below. The period of the rhythm is sensitive to supplemental fatty acids in the growth medium, although the period of the wild-type is not affected by any supplemental fatty acid. The unsaturated fatty acids, 18:1, 18:2, and 18:3, were first found to alter the period of *cel*, lengthening it to as long as 26, 40 and 33 h, respectively, at 22°C (Brody and Martins 1979). The changes in period produced by 18:2 in *cel* interact in double mutant strains with the period changes caused by the *chr* (*chrono*) or *frq* mutations. The periods of the double mutants *cel; chr* and *cel; frq* demonstrate partial epistasis between the mutations (Lakin-Thomas and Brody 1985a; Lakin-Thomas et al. 1990), suggesting that 18:2 may, in *cel*, affect the same part of the clock as the *frq* and *chr* gene products.

Fig. 2.1. Pathways of phospholipid synthesis in lower eukaryotes. **a** Fatty acid synthesis is catalyzed by the fatty acid synthase complex (FAS), and the product (16:0) may be elongated and/or desaturated as the CoA derivative. Further desaturation of 18:1 utilizes a phospholipid (Ptd-X) as the substrate. The *cel* mutation is defective in the FAS complex. **b** Synthesis of phospholipids. In the presence of supplemental Etn or Cho, the CDP-dependent pathway for PtdEtn or PtdCho synthesis is active. PtdEtn and PtdCho may also be synthesized through a base-exchange pathway: CDP-DAG + X → Ptd-X + CMP. The *chol-1* mutant is defective in the first methylation step in the conversion of PtdEtn to PtdCho. Abbreviations: CDP, cytidine-5'-diphosphate; Cho, choline; CL, cardiolipin; CMP, cytidine-5'-monophosphate; CoA, coenzyme A; CTP, cytidine-5'-triphosphate; Etn, ethanolamine; Me, methyl group (donated by *S*-adenosylmethionine); Pi, phosphate; PtdEtn-Me, phosphatidylmonomethylethanolamine; PtdEtn-diMe, phosphatidyldimethylethanolamine; PtdGly, phosphatidylglycerol; PtdGlyP, phosphatidylglycerol phosphate; PtdSer, phosphatidylserine.

a

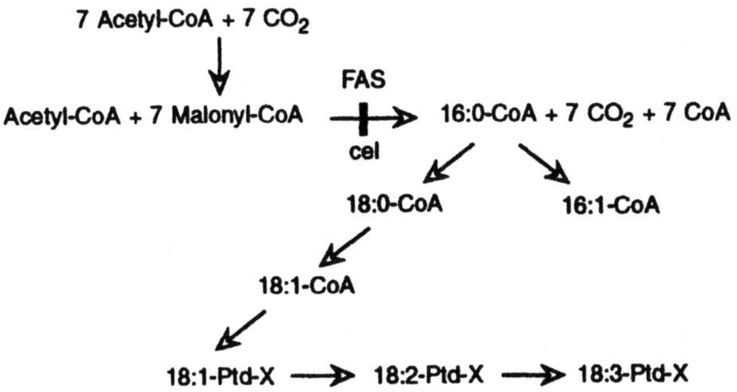
7 Acetyl-CoA + 7 CO2
FAS
Acetyl-CoA + 7 Malonyl-CoA
16:0-CoA + 7 CO2 + 7 CoA
cel
18:0-CoA
16:1-CoA
18:1-CoA
18:1-Ptd-X
18:2-Ptd-X
18:3-Ptd-X

b

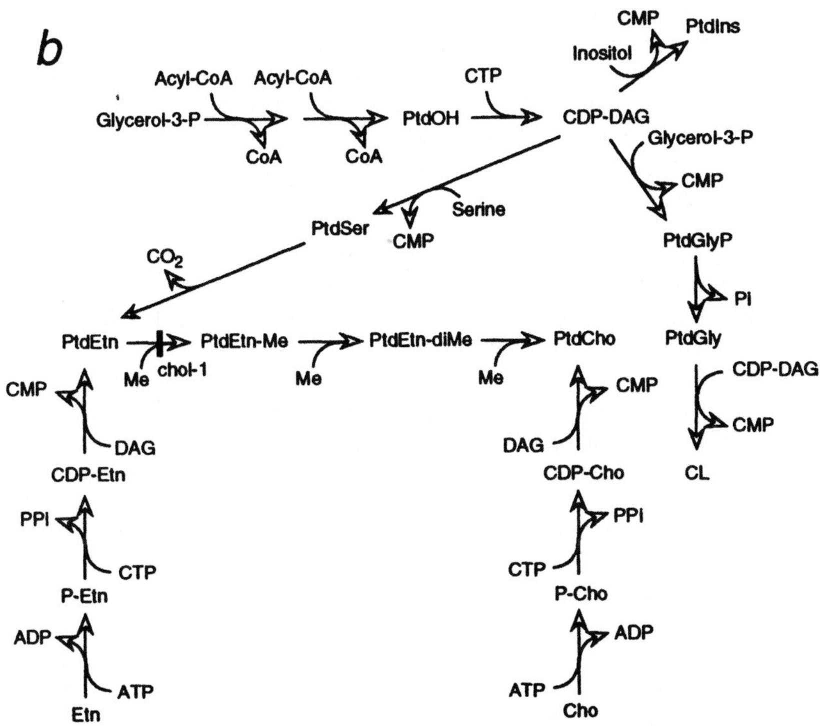
CMP
PtdIns
Inositol
Acyl-CoA
Acyl-CoA
CTP
Glycerol-3-P
PtdOH
CDP-DAG
CoA
CoA
Glycerol-3-P
CMP
Serine
PtdSer
CMP
PtdGlyP
CO2
PI
PtdEtn
PtdEtn-Me
PtdEtn-diMe
PtdCho
PtdGly
Me
chol-1
Me
Me
CDP-DAG
CMP
CMP
CMP
DAG
DAG
CDP-Etn
CDP-Cho
CL
PPI
PPI
CTP
CTP
P-Etn
P-Cho
ADP
ADP
ATP
ATP
Etn
Cho

Effects of Fatty Acids on the Period. A variety of fatty acids, both natural and synthetic, have now been tested for ability to lengthen the period of the *cel* mutant at 22°C (for review, see Lakin-Thomas et al. 1990). Saturated fatty acids 8 to 13 carbons long lengthen the period, while shorter (6 carbons) or longer (14 to 24 carbons) do not. The unsaturated fatty acids, 18:1, 18:2, and 18:3, which occur naturally in *Neurospora*, and $\Delta^{6,9,12}$-18:3 (γ-linolenic acid), which does not, all lengthen the period. Most monounsaturated fatty acids tested do not. A *cis* double bond is not necessary for period lengthening since *trans, trans*-$\Delta^{9,12}$-18:2 lengthens the period, as does the triple bond analog of 18:1 (9-octadecynoic acid). Very unusual analogs in which the double bonds of 18:1, 18:2, or 18:3 have been replaced variously with bromine, epoxy, methylene, or alkoxy substituents have also been tested. Some lengthen the period while others do not, and there appears to be no simple way to predict the effect of any given analog. In contrast to these results, the fatty acids 14:0 (myristic acid) and 16:0 favor normal circadian periods and reverse the effects of other fatty acids. At least some of the fatty acids act by altering temperature compensation, as discussed below. Fatty acids may also affect phase resetting properties; with supplemental 18:2, *cel* strains are less responsive to phase resettting by light than is *cel*$^{+}$ (S. Brody, unpublished).

Possible Mechanisms of Action of Fatty Acids. It is not known how fatty acids lengthen the period of the *cel* mutant, or, indeed, if they all act in the same way. Three general mechanisms can be suggested by which fatty acids might act: fatty acids could affect membrane properties; they could affect specific proteins, possibly by acting as second messenger molecules; or they could affect general cell metabolism. Here, we will focus specifically on the possibility that exogenous fatty acids act by altering the membrane lipids of the *cel* mutant. It must be emphasized, however, that mechanisms in which membrane lipids are unaffected have not been ruled out. All three general mechanisms by which fatty acids might work are discussed in detail by Lakin-Thomas et al. (1990).

If fatty acids act by perturbing the membranes of *cel* the wild-type must somehow be protected from such perturbation. As there is no evidence that any protein other than the fatty acid synthase is altered in *cel*, this protein apparently must play an integral role in the sensitivity of membranes, and hence of the rhythm to fatty acids. Presumably, reduced activity of this enzyme leaves the strain unable to properly regulate metabolism, either of endogenous membrane lipids, of added lipids, or of both.

The most obvious way that fatty acids could act on membranes is by being incorporated into the lipid bilayer and altering the properties of the membrane. Fatty acids could either directly partition into the bilayer or they could be incorporated into membrane phospholipids. Presumably the wild-type is not affected by these same fatty acids because, having no deficiency in fatty acid synthesis, it does not incorporate supplemental fatty acids into phospholipids or otherwise allow them to enter membranes to the same extent as the mutant. We have shown, using radiolabeled fatty acids, that the *cel* mutant both oxidizes and incorporates into its mycelia more supplemental fatty acids than does the wild-type (Coté 1986).

Many of the fatty acids which lengthen the period in *cel* are known to disrupt membranes and increase their fluidity when incorporated into membrane lipids, including *cis*-unsaturated fatty acids, short-chain saturated fatty acids, and fatty acids with bulky alkoxy side chains. This suggests that increased membrane fluidity could lengthen the period. Certain fatty acids which should increase fluidity, including positional isomers of 18:1 and certain analogs with bulky side chains, do not lengthen the period; however, these fatty acids might not be incorporated into membranes in *cel*, or might be rapidly converted to derivatives which do not increase fluidity. On the other hand, *trans, trans*-18:2 does not increase fluidity but it does lengthen the period. It is difficult to envision metabolism which could convert this fatty acid into a membrane-disruptive fatty acid, but such metabolism has not been ruled out.

The uptake of saturated fatty acids 6 to 24 carbons long into mycelia of the *cel* mutant was measured by gas chromatography of total fatty acids extracted from the mycelia of *cel* cultures grown with the fatty acid supplement (Mattern and Brody 1979). This procedure does not distinguish between free fatty acids and those incorporated into lipids. Only supplemental fatty acids which do not occur naturally in *Neurospora* lipids could be analyzed, and, of these, only 17:0 and 20:0 (arachidic acid), which do not lengthen the period, could be detected in the mycelia. Furthermore, ^{14}C-labeled 12:0 (lauric acid), a period lengthening fatty acid, was rapidly taken up by the *cel* strain, but was mostly oxidized, the label being released as radioactive carbon dioxide (Mattern and Brody 1979). Further evidence against fatty acids acting by incorporation into lipids comes from experiments with fatty acids having methoxy side chains. While only the monomethoxy derivative lengthened the period, the dimethoxy and trimethoxy derivatives were incorporated into the lipids of the *cel* strain to an extent five to nine times greater than the monomethoxy derivative (Mattern 1985a,b). Based on these and other observations on the incorporation of fatty acids into *cel*, Mattern (1985a,b; Mattern and Brody 1979) has rejected the hypothesis that fatty acids lengthen the period by directly altering the bulk fluidity of cellular membranes. However, we cannot rule out the possibility that the fatty acids might partition into, or be incorporated into the lipids of specific membranes, and thereby specifically fluidize these membranes.

Another way fatty acids could affect *cel* is by altering the regulation of membrane lipid composition. One could postulate that the defect in *cel* in fatty acid synthesis upsets normal regulation of lipid synthesis, leading to an inability to maintain normal membrane lipid composition, and that exogenous fatty acids further perturb this regulation. Changes in membrane composition would in turn affect rhythmicity if membrane lipids are part of or intimately associated with the clock mechanism. It is not necessary that the fatty acids enter the membranes in this hypothesis, only that they affect regulation of membrane composition. *Neurospora* mitochondria may possess a fatty-acid synthesizing system distinct from the cytoplasmic fatty acid synthase (Mikolajczyk and Brody 1990). As one would expect mitochondrial fatty acid synthesis to be regulated differently than cytosolic fatty acid synthesis this might explain why the clock of *cel* is altered. If the mitochondrial system is forced to take over a significant part of fatty acid synthesis in this mutant it would be expected to produce alterations in the regulation of membrane composition compared to the wild-type.

Fatty Acid Composition. The phospholipid fatty acid composition of the *cel* mutant and of the wild-type have been examined in liquid culture at 22°C. The phospholipid fatty acid composition of both strains varies during logarithmic growth under these conditions, but the phospholipid fatty acid composition of *cel* varies in a way much different than that of cel^+. Notably, in the wild-type the level of 18:2 increases with growth and the level of 16:0 is constant, while in *cel* the level of 18:2 is constant and the level of 16:0 increases with growth .

Exogenous fatty acids have no effect on the temporal pattern of phospholipid fatty acid composition in the wild-type, but both supplemental 16:0 and supplemental 18:2 alter the temporal pattern of *cel* (Coté and Brody 1987b). Surprisingly, the level of 16:0 in 16:0-supplemented cultures of *cel* is less than in unsupplemented cultures, as if *cel* overcompensated for the presence of the exogenous fatty acid. Both 16:0 and 18:2 alter the phospholipid fatty acid composition pattern in a similar manner, making it more like the pattern of the wild-type; however, 18:2, which lengthens the period of *cel*, makes the phospholipid fatty acid composition pattern of *cel* even more like that of the wild-type than does 16:0, which does not lengthen the period. Thus, while regulation of membrane fatty acid composition is clearly altered in *cel*, the observed changes in fatty acid composition cannot explain the different effects of these fatty acids on period. It should be noted, however, that only fatty acids up to eighteen carbons long were analyzed in these studies. It is not known whether the levels of longer fatty acids, which are very minor constituents of wild-type *Neurospora* membranes, are significantly altered in the *cel* mutant.

Evidence for altered membrane composition in *cel* in these experiments may not be relevant to the defects in this strain's circadian rhythm. The measurements were of cultures grown in liquid medium in constant light, where rhythmicity is not expressed. Phospholipid fatty acid composition has, therefore, also been measured in the young mycelium just behind the growing front of both wild-type and *cel* cultures on solid medium (Coté and Brody 1987b). Again, the *cel* strain, but not the wild-type, appears to overcompensate for exogenous 16:0 and has less of this fatty acid when grown on solid medium supplemented with it. With either 16:0 or 18:2 supplement, the 18:2 level was significantly higher in *cel* than in cel^+ while the 16:0 level was significantly lower, thus, again, fatty acids perturb the bulk membrane composition of *cel* but this does not correlate with their period-lengthening ability. Furthermore, it has not been demonstrated that rhythmicity persists in older areas of *cel* cultures on agar, as it does in the wild-type, or that the period of the rhythm in older areas is affected by fatty acids. Thus, fatty acids are known to alter rhythmicity only in the narrow strip of mycelium at the growing front where the banding pattern is laid down. The fatty acid composition of this narrow strip has not been measured.

Mitochondrial Metabolism. The phospholipid fatty acid composition of the mitochondria of *cel* and the wild-type were also examined in cultures in liquid medium (Coté and Brody 1987b). Each strain had a fatty acid composition pattern in its mitochondrial phospholipids essentially identical to that of the bulk mycelial phospholipids; in other words, the *cel* strain is altered in mitochondrial membrane fatty acid composition compared to the wild-type. Studies of the effect of carbon source on rhythmicity in *cel* implicate the inner mitochondrial membrane in period lengthening induced by fatty acids. Carbon sources which cannot be fermented and thus must be metabolized in the mitochondria — acetate, glycerol, glutamate, casamino acids, pyruvate, and ethanol — reverse the period-lengthening effect of supplemental fatty acids (Coté 1986). Tween 40 and tween 80, when provided as sole carbon sources, also reverse period lengthening. These compounds are water-soluble esters of fatty acids; tween 40 is an ester of 16:0 and tween 80 of 18:1. Only the fatty acid portions of these compounds are metabolized — by ß-oxidation in the mitochondria. Conversely, none of the sugars tested, including both pentoses and hexoses, both monosaccharides and disaccharides, and both ketoses and aldoses, reversed fatty acid-induced period lengthening.

In *Neurospora* growing on nonfermentable carbon sources we would expect, relative to growth on sugars, a shutdown of glycolysis, and increased activities of gluconeogenesis, the glyoxylate cycle, and the pentose phosphate pathway. We would also expect increased levels of mitochondrial enzymes and much increased flux through mitochondrial energy-yielding pathways. Electron flow through the respiratory electron transport chain would be increased, and possibly also through the alternate, cyanide-insensitive electron transport chain. The possibility that the metabolic change relevant to rhythmicity may be mitochondrial is supported by the finding that antimycin A, an inhibitor of mitochondrial cytochrome *c* reductase, also reverses period lengthening by fatty acids (Brody et al. 1987).

Two genetic loci are known at which mutations both alter mitochondrial metabolism and reverse period lengthening by fatty acids in the *cel* mutant: the maternally inherited locus (two alleles, [*mi-2*] and [*mi-5*]) (Brody et al. 1987) and the oligomycin resistance (*oli*^{r}) locus (Lakin-Thomas and Brody 1985a). The [*mi-2*] and [*mi-5*] mutations are mutations of the mitochondrial genome which lead to defects in subunit 1 of the mitochondrial cytochrome *c* oxidase. Mutations at the *oli*^{r} locus alter subunit 9 of the mitochondrial ATP synthase (the DCCD-binding protein), conferring oligomycin resistance. Several alleles of this mutation are known and their ability to reverse period lengthening correlates with the degree of their oligomycin resistance (Lakin-Thomas and Brody 1985a). It is interesting that 18:2, one of the fatty acids whose period lengthening is reversed by the *oli*^{r} mutation, itself confers a modest degree of oligomycin resistance (Brody and Forman 1980).

The *oli*^{r}, [*mi-2*], and [*mi-5*] mutations, and continuous exposure to antimycin A, all of which reverse fatty acid effects, increase levels of mitochondrial protein relative to total protein in the wild-type (Brody et al. 1989; Lakin-Thomas et al. 1990), consistent with the possibility that they affect the *cel* mutant by altering mitochondrial metabolism. Interestingly, they also shorten the period of the wild-type to 18 to 20 h (reviewed in Lakin-Thomas et al. 1990); thus altered mitochondrial metabolism may also affect rhythmicity in the wild-type, although to a lesser extent than in the *cel* mutant.

2.3.2 Fatty Acid Composition of Clock-Affecting Mutants

The altered temporal pattern of phospholipid fatty acid composition in *cel* led us to examine the phospholipid fatty acid compositions of other mutants in which rhythmicity was altered (Coté and Brody 1987a). We reasoned that while drugs and environmental factors might produce adaptive changes in membrane composition which preserve membrane properties, genetic defects might produce non-adaptive changes which alter membrane properties and affect cellular processes dependent on them, including perhaps circadian rhythms. Bulk phospho-lipid fatty acid composition has a normal temporal pattern in mutants at the *chr* and *frq* loci. The *frq* locus has alleles producing both long and short periods, some of which also affect temperature compensation. This suggests that altered period and temperature compensation can result from factors other than altered membrane fatty acid composition. It should be noted that although membrane fatty acid composition is normal in *frq* mutants we cannot say that their membrane composition is normal since the phospholipid species composition has not been investigated.

The pattern of bulk fatty acid composition was also examined in another clock-affecting mutant, the *prd-1* mutant (Coté and Brody 1987a), which has a period of 25.6 h. This mutant has a bulk phospholipid fatty acid composition different from both that of the wild-type and that of *cel*, both in liquid culture and on plates. Interestingly, the *prd-1* mutation also reverses period lengthening by fatty acids in the *prd-1; cel* double mutant (Lakin-Thomas and Brody 1985a). The *prd-1* mutant also apparently produces 2 to 2.5 times as much mitochondrial protein per unit mycelial mass as does the wild-type (Brody et al. 1989). This mutant thus resembles the [*mi*] and *oli*^{r} mutants which both reverse the effects of fatty acids on *cel* and increase levels of mitochondrial protein. It differs from them, however, in lengthening the period rather than shortening it. Strangely, however, while the fatty acid composition of the bulk phospholipids of *prd-1* is altered, that of the isolated mitochondria follows the same pattern as in the wild-type (Coté 1986).

2.3.3 The *chol-1* Mutant

The effects of inositol depletion on light sensitivity, and the effects of the *cel* mutation on the period, led to an examination of other *Neurospora* mutations defective in lipid metabolism, in particular *chol-1* (Lakin-Thomas 1995). The *chol-1* mutation is defective in the first methylation step in the synthesis of PtdCho from phosphatidylethanolamine, and requires choline, or the PtdCho precursors monomethylethanolamine or dimethylethanolamine, for normal growth (see Fig. 1b). On low levels of choline supplementation, the phospholipid headgroup composition is altered such that the levels of PtdEtn and PtdIns increase, and the level of PtdCho decreases (Hubbard and Brody 1975; Juretic 1977). Choline depletion also results in an accumulation of neutral lipids (Juretic 1977) and a decrease in the ratio of phospholipid to protein in the mitochondria (Luck 1964). This strain is "leaky" in that it will grow on minimal medium without choline at a rate about half that on high choline medium. There is some residual methylation activity (Scarborough and Nyc 1967), and an appreciable amount of PtdCho is present in strains grown without choline supplementation (P.L. Lakin-Thomas, unpublished results).

A *chol-2* mutant, defective in the second methylation step in PtdCho synthesis, has also been isolated, and the effects of choline depletion on lipid metabolism in the *chol-1*; *chol-2* double mutant strain are described below under "Mutants Defective in Temperature Compensation." It is not known whether the effects of choline depletion on the period of *chol-1* described in this section also apply to the *chol-1*; *chol-2* strain.

On high levels of choline, the *chol-1* mutant is similar to the wild type with respect to period and growth rate. On low levels of choline and on minimal medium without choline the growth rate is slower and the period is longer than wild-type, the period reaching a maximum of about 58 h at 22°C in the absence of choline. Below about 10 μM choline, there is an inverse linear relationship between period and growth rate. The period of choline-depleted cultures is variable within each race tube and, as a result, replicate cultures fail to remain in phase with each other; thus, choline depletion appears to disrupt phase control as well as lengthening the period.

Choline depletion of the *chol-1* strain does not abolish sensitivity to light, and the rhythm in these cultures can be damped out by continuous light. Choline-depleted cultures can also be entrained to light/dark (LD) cycles close to their intrinsic period; for example, cultures on minimal medium, with an intrinsic period of about 58 h, can be entrained to a 48 h LD cycle, but will not entrain to a 24 h LD cycle. These results indicate that the light-sensitive oscillator is still functioning in choline-depleted cultures.

PtdCho is one of the most abundant phospholipids in *Neurospora*, and depletion of this major constituent of membranes could have many effects on clock components. Preliminary experiments with cultures labeled to isotopic equilibrium during growth with [^{3}H]16:0 and various levels of choline indicate that the levels of label in PtdCho and in other phospholipids do not correlate with period, but that the levels of label in DAG and in other neutral lipids show linear correlations with period (P.L. Lakin-Thomas, unpublished results). It may be that it is not an alteration in the phospholipid composition of the membrane that is responsible for the lengthened period in choline-depleted cultures, but rather an excess of free fatty acids and DAG, known to be activators of protein kinases and regulators of many cellular processes (Billah and Anthes 1990; Graber et al. 1994).

2.3.4 Other Lipid Mutants

Do all mutations that affect lipid synthesis also affect the period? The *inl* mutation, which under some conditions affects light sensitivity, is an obvious candidate to test this hypothesis. Unfortunately, this mutation fails to conidiate on solid medium with low levels of inositol, and it is therefore impossible to assay the period under these conditions. Four other mutants (reviewed in Lakin-

Thomas et al. 1990) are relevant to this question. The *phen-1* (phenylalanine requiring) mutant, which may be blocked in the synthesis of certain sterols, has a period of 19 h on high levels of phenylalanine. Three other mutants have no effect on the period: two mutants deficient in ergosterol synthesis, *erg-1* and *erg-3* and an unsaturated fatty acid-requiring mutant, *ufa-1*, blocked in the desaturation of stearic acid. These last three mutants were tested only on levels of the appropriate supplement sufficient to allow normal growth, and therefore the effects of ergosterol or unsaturated fatty acid depletion are not known.

2.4 Lipids and Output

Rhythms in various aspects of lipid composition and metabolism have been observed in a variety of organisms from unicellular algae to vertebrates. In some cases, rhythms have been monitored only in entrained organisms; these rhythms might be driven either by an endogenous oscillator or by the periodic environmental stimuli entraining the organism. In other cases, rhythms have been shown to persist under constant conditions and thus must be driven by an endogenous oscillator; they may either be an expression of an output driven by the oscillator or part of the oscillator mechanism itself. Since there is no compelling evidence so far to indicate that any lipid rhythm is part of the oscillator, it is conservative to treat these rhythms as part of an output mechanism.

A rhythm of fatty acid composition has been reported in *Neurospora* (Roeder et al. 1982), and this rhythm has been studied in considerable detail. The rhythm, an oscillation in the mole percent of the two major unsaturated fatty acids of *Neurospora*, 18:2 and 18:3, was seen both in total lipids and in phospholipids from cultures of *bd csp-1* grown on solid medium. (The *csp-1* mutation inhibits conidial release and simplifies laboratory manipulation.) The rhythms in these two fatty acids were approximately 180° out of phase with each other, suggesting possible rhythmic control of the conversion of 18:2 to 18:3. These rhythms were also found in the *bd*$^{+}$ *csp-1* strain which shows little, if any, rhythmic conidiation. The fatty acid rhythms are therefore apparently not simply part of the conidiation process. It is not know if the rhythm in fatty acid composition persists in rhythmic liquid cultures, which also have no obvious conidiation rhythm.

These observations were extended by three additional studies. First, individual phospholipid head group composition was analyzed as a function of circadian phase but no significant oscillations were found (P. Roeder and S. Brody, unpublished). This is the only attempt to find a rhythm of head group composition in a free-running organism that we know of. Second, fatty acids in the *sn-1* and *sn-2* positions of total phospholipids were differentiated by using phospholipase A_2 to release the fatty acids in the *sn-2* positions. Oscillations in 18:2 and 18:3 were found in both *sn-1* and *sn-2* positions, with approximately equal amplitude (M. Adamich and S. Brody, unpublished). Third, fatty acids in individual phospholipids were assayed in extracts from different circadian phases. Rhythms in 18:2 and 18:3 were found in both major phospholipids of *Neurospora*, PtdCho and PtdEtn (P. Roeder and S. Brody, unpublished). The amounts of the other phospholipids extracted were too low to accurately determine their fatty acid composition. These additional studies indicate that the oscillations in 18:2 and 18:3 are very general, not being confined to a particular type of lipid or a stereospecific position in the lipids. It is not yet known if these fatty acid oscillations occur in membranes of all cellular organelles of *Neurospora*.

Rhythms in fatty acid composition and desaturation have also been reported in plants and animals, and may represent a general phenomenon. In *Acetabularia mediterranea*, 18:1 levels oscillated in whole cell lipids and 16:0 oscillated in chloroplast lipids (Jerebzoff and Vanden Driessche 1983). In developing spinach leaves oscillations in 18:1 and 18:3 levels could be detected in total lipids and in PtdCho in particular (Browse et al. 1981). These rhythms were attributed to inhibition of 18:1 synthesis in the dark, while 18:1 and 18:2 desaturation continued. A rhythm in the desaturation of exogenous radioactively labeled 18:1 and 18:2 by cucumber seedlings was also reported (Murphy and Stumpf 1980). The iodine number, which indicates the level of fatty acid unsaturation, has been reported to oscillate in rat brain lipids (Díaz-Muñoz et al. 1985). In rat liver, diurnal variations in fatty acid composition and in microsomal desaturase activity have been found (Actis Dato et al. 1973; de Gomez Dumm et al. 1984). None of these rhythms were studied under constant conditions, and so may not be driven by an endogenous oscillator. A fat-free diet eliminated the fatty acid rhythm in rat liver and altered the desaturase rhythm, indicating that the fatty acid rhythm followed the rats' feeding cycle and not an endogenous oscillator.

Circadian variations in the fatty acid composition of phospholipids under constant conditions have also been observed in a number of organisms. Adamich (1976) reported circadian rhythms in the levels of the various polyunsaturated fatty acids in *Gonyaulax polyedra*. Lecharny and coworkers (1990) found that 18:1 levels in PtdCho and in PtdEtn varied with a circadian rhythm in growing stems of *Chenopodium rubrum*. Gardner (1979; Gardner and Galston 1977) found circadian rhythms in 16:0 and 18:2 levels and in the ratio of saturated to unsaturated fatty acids in *Phaseolus coccineus* pulvini. Rikin and coworkers (1993) reported rhythms in the levels of 18:2 and 18:3 in cotton seedlings which correlated with a rhythm in chilling resistance. Both fatty acids peaked in the subjective night, but 18:2 levels oscillated in the membrane phospholipids while 18:3 levels oscillated in the membrane glycolipids.

2.5 Lipids and Temperature Compensation

2.5.1 Introduction

The term "temperature compensation" describes the conservation of the period of a circadian oscillator at different constant temperatures. Temperature effects in biology are often quantified using the concept of Q_{10}, which is defined as the ratio of the rate of a process at one temperature to its rate at a temperature 10 degrees lower. For a rhythmic process, the frequency is its rate: a Q_{10} of 2 means that the frequency doubles with a 10 degree increase in temperature. A mathematically equivalent definition is the ratio of the period at the lower temperature to the period at a temperature 10 degrees higher. Most biological and chemical processes have Q_{10}'s of 2 to 3, while most circadian rhythms have Q_{10}'s close to 1 and are therefore said to be temperature compensated.

Temperature compensation might be either an innate part of the oscillator mechanism itself, or an input pathway which uses a temperature sensor to adjust the frequency of the oscillator. There is as yet no definitive experimental evidence to distinguish between these two possibilities. The existence of mutations such as *cel*, which eliminate temperature compensation without eliminating rhythmicity *per se*, might suggest that temperature compensation is an input pathway separate from the oscillator itself. However, it could also be argued that temperature compensation is an intrinsic part of the oscillator, and mutations such as *cel* alter the temperature compensation properties of oscillator components without disrupting the oscillation entirely.

If temperature compensation is intrinsic to the oscillator then mutations that eliminate temperature compensation are likely to directly affect oscillator components. Aronson et al. (1994a) have suggested that temperature compensation is likely to be derived directly from the components of the oscillator or the manner in which the components are interconnected. Their conclusion is based on the phenotype of mutants at the *frq* locus which alter the period of the rhythm, abolish temperature compensation, and disrupt the normal stable expression of rhythmicity. They propose that the FRQ gene product is an intrinsic component of the oscillator, and therefore temperature compensation, disrupted by *frq* mutations, must also be intrinsic to the oscillator. If this argument is extended, then the loss of temperature compensation in *cel* and the demonstrated interactions between *cel* and *frq* (Lakin-Thomas and Brody 1985a) might indicate a central role for *cel* in the pacemaker as well.

The biochemical mechanism of temperature compensation is not known for any circadian oscillator, although a number of theories have been advanced to explain it (reviewed in Lakin-Thomas et al. 1990). In this review we will discuss evidence for a role for membrane lipids in maintaining temperature compensation of the rhythm. Organisms exposed to different ambient temperatures are known to adjust the lipid composition of their membranes to maintain membrane fluidity, and this compensation of fluidity could underlie temperature compensation of the rhythm. A model for the circadian oscillator based on this property of compensation of membrane fluidity was proposed by Njus et al. (1974), but little experimental evidence was offered in support of this model. The discovery that the *cel* mutation in *Neurospora* affects temperature compensation as well as fatty acid synthesis was an important finding in support of this model (Mattern et al. 1982). Furthermore, the *chol-1* mutant, which affects phospholipid synthesis, also affects temperature compensation (P.L. Lakin-Thomas, unpublished results). The *frq* locus, several alleles of which have altered temperature compensation or lack it altogether, show genetic interactions with the *cel* mutation in double mutant strains (Lakin-Thomas and Brody 1985a). In the following sections, we discuss the general relationship between temperature and lipids and the properties of mutations which abolish temperature compensation.

2.5.2 Temperature Effects on Membrane Lipid Composition

Temperature-induced changes in membrane lipid composition have been reviewed many times (Harwood 1983; Thompson 1983; Cossins and Raynard 1987, among others). If membrane composition did not change in response to temperature, an increase in temperature would increase membrane fluidity while a decrease in temperature would decrease membrane fluidity. To restore normal membrane fluidity, membrane composition might be modified in several ways. The fatty acid composition of the phospholipids or sphingolipids, the sterol content, the phospholipid head group composition, and the carbohydrate species attached to sphingolipids could all be altered. One generality is that in most species an increase in temperature leads to an increase in the relative amounts of saturated fatty acids, which decrease fluidity, while a decrease in temperature leads to an increase in the relative amounts of monounsaturated and polyunsaturated fatty acids, which increase fluidity. The details of this response differ from one organism to another.

Changes in lipid composition, that is lipid remodeling, in response to temperature changes occur in two stages, a relatively rapid response, occurring within minutes, and a long-term response occurring over a scale of hours. The rapid responses involve acyl transfers between molecules of specific types within the phospholipids, producing, for example, increased levels of di-unsaturated species of PtdCho (Lynch and Thompson 1984). Little or no change in the overall fatty acid composition occurs during this stage. Long-term adaptations involve *de novo* fatty acid synthesis and fatty acid desaturation. During this stage induction or repression of specific genes related to membrane remodeling may occur. How temperature changes control the expression of genes is not known, but Vigh et al. (1993) have recently shown that the physical state of the membrane itself can affect the expression of a gene coding for a fatty acid desaturase in the cyanobacterium *Synechocystis*. Maresca and Cossins (1993) discuss recent developments in the area of gene regulation during temperature acclimation.

The role of lipids as second messengers and precursors of second messengers has recently become clear (reviewed by Liscovitch and Cantley 1994). Whether temperature-induced changes in membrane composition might affect an organism's ability to produce lipid-generated second messengers is a question which might be well worth exploring.

The temperature-dependence of membrane lipid compensation has been well studied in *Neurospora*, where regulation of 18:3 synthesis appears to be the primary mechanism of temperature acclimation. Martin et al. (1981) examined the fatty acid composition of cultures grown at 37°C and shifted to 15°C. They found an increase in the mole percent of 18:3 in phospholipids which could not be accounted for by transfer of unsaturated fatty acids from the neutral lipids or by transfer of saturated fatty acids to the neutral lipids. Although the degree of unsaturation differed between lipid species at a given temperature, the changes in unsaturation in response to changes in temperature were similar in each lipid species examined (Martin et al. 1981). Vokt and Brody (1985) did the opposite experiment, monitoring changes in phospholipid fatty acid composition in cultures shifted from 22 to 40°C. The mole percent of 18:3 decreased, and that of 18:1 increased, but there was no change in the levels of 16:0 or 18:0 and only small changes in 18:2 levels. Cultures shifted to the higher temperature reached their new steady-state fatty acid composition within 6 hours of the temperature shift. The decrease in 18:3 was caused by dilution during growth and not by oxidation. Both Martin et al. (1981) and Vokt and Brody (1985) found that cultures grown continuously at various temperatures displayed a pattern of fatty acid composition consistent with the changes seen on temperature shifts. Levels of 18:1 were relatively elevated and levels of 18:3 relatively reduced at higher temperatures. Similar patterns were seen in the total lipid, neutral lipid, and phospholipid fractions (Vokt and Brody 1985) and in the plasma membrane (Aaronson and Martin 1983). In mitochondria, the fatty acid composition of PtdCho and PtdEtn changed with temperature while surprisingly the cardiolipin fatty acid composition changed very little (Schlame et al. 1993).

Changes in the relative proportions of lipid species have also been reported to play a role in temperature acclimation in *Neurospora*. In comparing cultures grown at 15 and 37°C, the same proportion of total fatty acids were found in the neutral lipid and phospholipid fractions (Martin et al. 1981), and the ratio of total sterol (free plus esterified) to phospholipid was constant (Aaronson et al. 1982). However, at the higher temperature the molar ratio of free sterol to phospholipid in whole cell lipids was almost double the ratio at the lower temperature (Aaronson et al. 1982). Despite this pattern in whole cell lipids, the free sterol to phospholipid ratios were lower at the higher temperature in mitochondrial and microsomal fractions (Aaronson et al. 1982) and in the plasma membrane (Aaronson and Martin 1983). The lower plasma membrane free sterol to phospholipid ratio resulted from an increase in the mole percent of phospholipid at the higher temperature. A corresponding decrease in the mole percent of glycosphingolipids at the higher temperature resulted in a constant ratio of sterol to total polar lipids (Aaronson and Martin 1983). The glycosphingolipid fraction of the plasma membrane was also found to vary in the proportion of long-chain hydroxylated fatty acids and in the long-chain base composition (Aaronson and Martin 1983). Changes in the proportions of different phospholipid species were also found at different growth temperatures: PtdCho levels decreased and PtdEtn levels increased at the lower temperature in total, mitochondrial, and microsomal lipids (Martin et al. 1981; Aaronson et al. 1982; Johnston et al. 1982), but in the plasma membrane, PtdEtn decreased and combined PtdSer plus PtdIns increased (Aaronson and Martin 1983).

Does temperature-induced remodeling of membrane composition contribute to temperature compensation of circadian rhythms? One way this could be tested is to examine the ability of *Neurospora* strains lacking temperature compensation to properly adjust their membrane lipid composition. The *cel* and *chol-1* mutants, in which lipid composition is known to be altered under at least some conditions, are prime candidates for such studies.

2.5.3 Mutants Defective in Temperature Compensation

The *cel* Mutant. The *cel* mutation and the *chol-1* mutation discussed below, are the only mutations affecting temperature compensation in which the primary biochemical defect is known. It is interesting that both these mutations also affect lipid synthesis. As discussed above, fatty acid synthesis is defective in *cel*. The mutation is leaky in that fatty acids are still synthesized, yet the altered temporal pattern of fatty acid composition of bulk phospholipids suggests that the normal regulation of membrane fatty acid composition is defective.

The effect of temperature on the period of the conidiation rhythm in the *cel* mutant is shown in Figure 2a. Temperature compensation is normal in this mutant at temperatures from 22 to 30°C; however, below about 22°C the Q_{10} is 2.2, and below 18°C it is 3.3 (Mattern et al. 1982). Thus, below a breakpoint temperature between 18 and 22°C the period of *cel* is even more temperature-sensitive than that of frq^9, discussed below. The fatty acids 18:1, 18:2, 18:3, and 12:0 (lauric acid) increase the breakpoint temperature to about 26°C with the result that the period is lengthened at 22°C (Mattern et al. 1982). Conversely, the fatty acids 14:0 (myristic acid) and 16:0 lower the breakpoint temperature to about 18°C, with the result that the period is normal down to this temperature. Acetate as a carbon source, like added 16:0 or 14:0, reverses the period-lengthening effect of fatty acids. It may also act by lowering the breakpoint temperature since the period of *cel* is circadian at 18°C on acetate medium (Coté, 1986). It is possible that all the period-lengthening and period-restoring effects reported for *cel* may result from changes in the pattern of temperature compensation in this strain, but this has not yet been demonstrated.

The *cel* mutant is temperature sensitive in another respect; at high temperatures, growth rate ceases to increase with temperature (Mattern et al. 1982). Like loss of temperature compensation, this phenomenon exhibits a breakpoint; below 23°C, the Q_{10} for growth is normal, approximately 2.6, while above 23°C it is 1.0. Interestingly, some of the fatty acids which alter the breakpoint temperature below which normal temperature compensation of rhythmicity is lost, also alter the breakpoint temperature above which normal temperature-dependence of growth rate is lost. The fatty acids 12:0 and 18:1, for example, reduce the temperature range of normal temperature responses by raising the breakpoint value for loss of temperature compensation, and lowering the breakpoint value for loss of growth-rate temperature dependence while the fatty acids, 14:0 and 16:0, broaden the range of normal temperature responses by lowering the breakpoint value for loss of temperature compensation and raising the breakpoint value for loss of growth-rate temperature dependence.

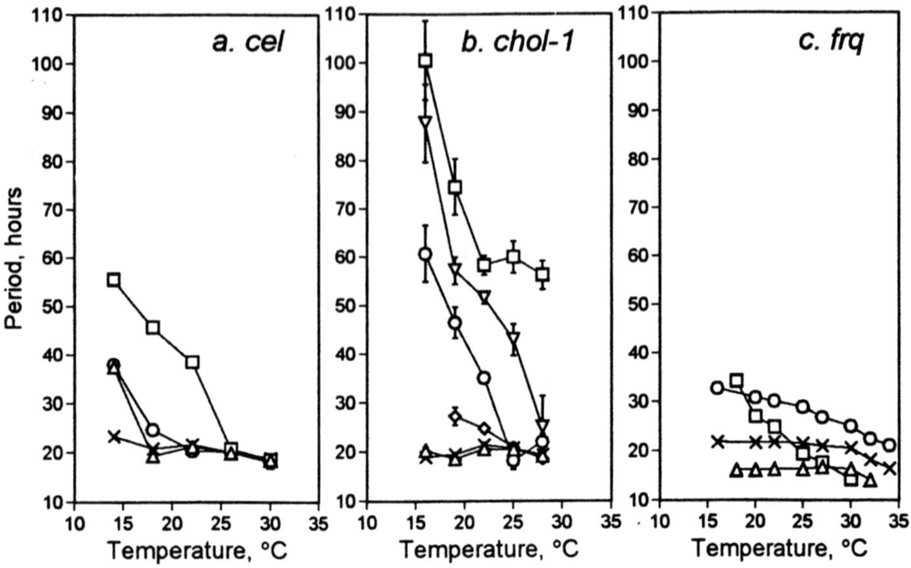

Fig. 2. Effects of mutations on temperature compensation of the conidiation rhythm in *Neurospora*. All strains carry the *bd* mutation. In **a** and **b**, all strains carry the *csp-1* mutation. All strains were assayed on maltose/arginine medium with the exception of frq^+, frq^1 and frq^7 in **c**, which were assayed on acetate medium. **a**: circles, *cel*, unsupplemented control; triangles, *cel*, 0.5 mM 14:0; squares, *cel*, 0.5 mM 18:2; crosses, cel^+, unsupplemented control. Data from (Mattern et al. 1982). SEMs ranged from 0.3 to 4.4 h, and averaged 1.2 h. **b**: squares, *chol-1*, minimal medium; triangles (down), *chol-1*, 0.5 µM choline; circles, *chol-1*, 2.5 µM choline; diamonds, *chol-1*, 10 µM choline; triangles (up), *chol-1*, 100 µM choline; crosses, $chol^+$, minimal medium. (From P.L. Lakin-Thomas, unpublished). Error bars indicate ± 1 SEM. **c**: triangles, frq^1; circles, frq^7; squares, frq^9, crosses frq^+. Data from Gardner and Feldman (1981), and Loros and Feldman (1986). SEMs for frq^+, frq^1 and frq^7 ranged from 0.4 to 0.6 hr, SEM for frq^9 was not reported.

The *chol-1* Mutant. Like the *cel* mutation, *chol-1* also affects temperature compensation, and the effect is dependent on the concentration of choline supplied in the medium as summarized in Figure 2b (P.L. Lakin-Thomas, unpublished results). At 20 µM choline and above, temperature compensation in *chol-1* is indistinguishable from that in wild-type, with a Q_{10} of about 1.0. On 10 µM choline, the frequency is less well-compensated between 19°C and 28°C (Q_{10} of 1.5). On 2.5 µM choline, period is overcompensated between 25°C and 28°C (Q_{10} of 0.5) and compensation fails below 25°C (Q_{10} of 3.8). Poor temperature compensation is found throughout the range 16°C to 28°C on 0.5 µM choline (Q_{10} of 2.9). Therefore, across this range of choline concentrations, choline depletion appears to increase the breakpoint at which temperature compensation is lost. Cultures on zero choline do not seem to fit this pattern:

between 22°C and 28°C, the period is long but compensation is good (Q_{10} of 1.1), and the period increases below 22°C (Q_{10} of 2.5).

The change in breakpoint temperature in *chol-1* is similar to the effect of supplemental fatty acids on the *cel* strain, as described above. Both strains exhibit Q_{10}'s of around 2.5 to 3.5 under conditions where temperature compensation is lost, and are therefore more sensitive to temperature than *frq*^{9}. The *cel* and *chol-1* strains are also similar in the loss of normal temperature-dependence of growth rate. Between 19°C and 28°C, the growth rate of *chol-1* without choline supplementation is approximately constant (P.L. Lakin-Thomas, unpublished results). These results may point to a common period-lengthening mechanism in both strains, perhaps related to the increase in neutral lipids produced by choline depletion of *chol-1*.

Can the effects of *chol-1* on temperature compensation be accounted for by alterations in temperature-induced remodelling of membrane composition? Johnston et al. (1982) looked at temperature acclimation in the *chol-1; chol-2* double mutant, which is defective in two of the methylation steps that convert PtdEtn to PtdCho. In comparison with a *chol*$^+$ strain, the choline-depleted double mutant strain had reduced levels of PtdCho and increased levels of PtdEtn and PtdIns. Wild-type cultures transferred from 37 to 15°C showed a small decrease in the mole percent of PtdCho and a small increase in PtdEtn, as described above. In the choline-depleted strain transferred from 37 to 15°C however, the level of PtdCho increased, and the levels of both PtdEtn and PtdIns decreased, with the overall result that the phospholipid composition of the choline-depleted strain was more like that of the wild-type at the lower temperature. In spite of large differences in phospholipid headgroup composition, the total fatty acid composition of the choline-depleted strain was identical to that of the wild-type at both growth temperatures. Because fatty acid composition differs between phospholipid species, with PtdCho the most highly unsaturated, the change in phospholipid headgroup composition in choline-depleted strains resulted in changes in the fatty acid composition of the individual phospholipids (Martin and Johnston 1983). Changes in the fatty acid composition of cardiolipin were the most dramatic in response to choline depletion, and unlike in wild-type cultures where changes in temperature result in rather small changes in cardiolipin fatty acid composition, temperature acclimation in choline-depleted cultures resulted in large changes in cardiolipin fatty acid composition. Membrane lipid fluidity in choline-depleted and wild-type cultures was assayed by fluorescence polarization in lipid vesicles, and choline depletion was found to decrease fluidity at both growth temperatures. The decrease in fluidity induced by choline depletion was comparable to the decrease induced by a 22°C temperature rise (Martin and Johnston 1983). Although the picture is far from simple, choline depletion does appear to disrupt the normal mechanisms for maintenance of membrane fluidity in reponse to temperature changes, at least in the *chol-1; chol-2* double mutant. Whether the *chol-1* mutation alone can produce similar changes in the adjustment of membrane composition is not known.

The *frq* Locus. Mutations at the *frq* locus, isolated and extensively characterized by Feldman and coworkers (1982), have had an important impact on thinking and experimental approaches in the field of *Neurospora* circadian rhythms. The *frq* mutations have clearly indicated the key role that the FRQ gene product plays; it is "kinetically poised" in the oscillator mechanism such that mutations which alter it can either speed up or slow down the clock mechanism. The point mutations at the *frq* locus are allelic, as evidenced by the gene structure (Aronson et al. 1994a). Rhythmic *frq* gene expression, and the effects of ectopic expression of *frq* on rhythmicity and phase, have recently been reported (Aronson et al. 1994b).

Temperature compensation in *frq* mutants has been reviewed (Lakin-Thomas et al. 1990) and Figure 2c illustrates temperature compensation properties of some of the *frq* alleles. While the short period allele frq^1 has no effect on temperature compensation, the long period allele frq^7 produces a partial loss of temperature compensation: its Q_{10} for rhythmicity is about 1.3 across the temperature range of 16 to 34°C (Gardner and Feldman 1981). The frq^9 allele, a nonsense mutation in the middle of the gene, produces apparent arrhythmicity during early growth, although a conidiation rhythm often eventually appears with a Q_{10} of about 2 (Loros and Feldman 1986; Aronson et al. 1994a). Not all replicate cultures, however, will show rhythmicity and the period of this strain is unusually variable between replicate cultures compared to the wild-type (Aronson et al. 1994a). Disruption of the *frq* gene by transformation with a foreign DNA fragment produces a strain, frq^{10}, with a phenotype identical to frq^9 (Aronson et al. 1994a). Null mutations at the *frq* locus, therefore, clearly abolish temperature compensation; rhythmicity *per se*, however, persists. Rhythmicity during early growth can be restored to the frq^{10} strain by the addition of certain fatty acids (S. Brody, unpublished results). Rhythmicity is also seen in *chol-1; frq*^{10} double mutant strains grown on low choline (S. Brody and P.L. Lakin-Thomas, unpublished results). Rhythmicity in the *frq* null mutants seems therefore to depend on growth conditions, and perhaps frq^9 and frq^{10} could be called conditionally arrhythmic mutations.

A connection between the FRQ gene product and membrane lipids is suggested by studies on *cel; frq* double mutant strains (Lakin-Thomas and Brody 1985a). The effects of the *cel* and *frq* mutations on the rhythm were not independent, nor were they completely epistatic (i.e., the phenotype of the double mutant being identical to that of one of the single mutants). Rather, the two genes interacted in a manner described as "partial epistasis" (Lakin-Thomas et al. 1990). For example, the periods of *cel* and frq^1 grown on 18:2 supplementation were 42.3 and 13.9 h, respectively; the double mutant *cel; frq*^{1} would have a predicted period of 29.3 h if both mutations affected the period independently, but the observed period was 34.2 h. This interaction suggests a connection between the function of the FRQ gene product and lipid metabolism in *Neurospora*. The membrane fatty acid composition of *frq* mutants is normal, as noted above; however, the phospholipid head group composition has not been studied, nor has the lipid composition of *cel; frq* double mutants been analyzed.

2.6 Conclusion

2.6.1 Membrane Lipids and Rhythmicity

Membrane lipids are of considerable importance in cellular metabolism. Signal transduction systems controlling cell physiology are membrane-localized, as are transport proteins involved in nutrient absorption and ionic homeostasis, and the proteins involved in energy production in mitochondria and chloroplasts. All of these are sensitive to the physical properties of the membranes in which they sit, and these properties in turn are dependent on membrane lipid composition. The composition of membranes can be adjusted to minimize changes in membrane properties, not only changes caused by changes in temperature as described above, but also changes caused by such other environmental factors as pressure, salinity, and chemical insults. Membrane lipids also serve as storage forms of second messengers and second messenger precursors, such as fatty acids, PtdOH, DAG, and Ins(1,4,5)P_3.

Given the importance of membrane lipids in cell physiology, a role for them in rhythmicity has long been suspected. Suggestive, but inconclusive, evidence has supported such a role, as noted in the Introduction. A few workers have reported negative evidence, but again this has been inconclusive. For example, several phenyl-substituted alcohols alter membrane fatty acid composition of *Neurospora*, yet only one, 2-phenylethanol, apparently affects circadian rhythmicity (Nakashima 1983). One could argue that the insensitivity of clock function to perturbation of membrane composition by these compounds precludes membrane involvement in clock function; however, one could also argue that the observed changes in membrane composition protect membrane properties from perturbation by the alcohols so that membrane properties and circadian rhythmicity are unaltered. Other workers (Scholübbers et al. 1984) found that age- and temperature-dependent changes in fluorescent polarization values of *Gonyaulax polyedra* labeled with a fluorescent dye showed no correlation with changes in the period or phase of the circadian bioluminescence rhythm. However, as the authors themselves point out, it is not clear what membrane properties are reflected by the fluorescent polarization values of intact *Gonyaulax* cells.

What is needed to test the involvement of membranes in rhythmicity is a system in which the normal adjustment of membrane composition to environmental parameters can be first characterized in detail and then disrupted by genetic or pharmacological manipulation. The ascomycete fungus *Neurospora* is particularly suited to such studies. It is amenable to detailed biochemical study, and a large number of mutants are available to disrupt its normal metabolism. It is not surprising that the most important data supporting a role of membrane lipids in rhythmicity comes from studies on *Neurospora*. Table 4 summarizes many of the studies which have been performed on *Neurospora* membrane lipid composition, including studies on rhythms in membrane lipid composition as well as studies correlating altered lipid metabolism with changes in rhythmicity. A number of mutants of *Neurospora* are altered in both circadian rhythmicity and in lipid metabolism (see Table 1), demonstrating a correlation, in *Neurospora* at least, between lipid metabolism and the mechanism of the circadian clock.

2.6.2 Lipids and Output

It is not surprising that many aspects of lipid metabolism show diurnal fluctuations in many organisms. Some of these may be directly driven by light/dark cycles or feeding cycles, but others clearly follow the endogenous circadian oscillator. Most or all of these rhythms may be output rhythms, that is, the result of temporal control of lipid metabolism by the clock.

It is possible that some lipid rhythms represent components of the circadian oscillator mechanism. Particularly intriguing in this respect are rhythms in membrane fatty acid composition which have been reported in a number of plant species, in *Acetabularia mediterranea*, in rat liver and brain, and in *Neurospora*. Although the details of these rhythms vary from species to species they could represent rhythms in similar membrane properties. However, in none of these species is there yet evidence that the fatty acid rhythms are actually part of the oscillator mechanism.

2.6.3 Lipids and Input

Many biological photoreceptors are membrane localized, and the as-yet unidentified blue-light photoreceptor of *Neurospora* is likely to fall in this category. Thus light signaling to the oscillator is likely to be sensitive to membrane composition. A direct role for membrane lipids in light signaling, as precursors of second messenger molecules, would not be surprising, at least in some species. In *Neurospora*, however, the evidence appears to rule out light signal transduction through production of Ins(1,4,5)P_3 or DAG by phospholipase C-catalyzed cleavage of PtdIns, or through production of PtdOH or DAG by phospholipase D-catalyzed cleavage of PtdCho, or through production of fatty acids or lysolipids by phospholipase A.

The mechanism or mechanisms by which temperature changes are sensed by the oscillator remain unknown. A role for temperature-induced changes in membrane properties in such a mechanism seems plausible, but there is, as yet, no evidence for or against such a role.

Table 4. Summary of studies on the interaction between rhythmicity and lipid metabolism in *Neurospora*

I. Rhythms in lipid metabolism
- A. Rhythmic variables
  - 18:2 and 18:3 levels in total lipids and phospholipids of *bd*$^+$ and *bd*[a]
- B. Non-rhythmic variables
  - Phospholipid head group composition in *bd*[b]
- C. Not tested
  - Sphingolipid, sterol, or neutral lipid composition
  - Membrane composition of specific organelles

II. Lipid metabolism in clock-affecting mutant strains
- A. Altered lipid metabolism
  - Total phospholipid fatty acid composition in *cel* and *prd-1*[c,d]
  - Mitochondrial phospholipid fatty acid composition in *cel*[d]
  - Phospholipid head group composition in *chol-1*[e,f]
  - Ratio of neutral lipids to phospholipids in *chol-1*[f,g]
  - Ratio of lipid to protein in *chol-1* mitochondria[h]
- B. Normal lipid metabolism
  - Phospholipid fatty acid composition in *frq* strains[c,i]
  - Mitochondrial phospholipid fatty acid composition in *prd-1*[j]
- C. Not tested
  - Levels of minor fatty acids, including very long chain fatty acids (>18 carbons) in *cel* or other mutants
  - Sphingolipid, sterol, or neutral lipid composition in clock-affecting mutants
  - Temperature-induced modification of fatty acid or phospholipid head group composition in mutants defective in temperature compensation
  - Fatty acid composition of organelles other than mitochondria

[a]Roeder et al. 1982.
[b]P. Roeder and S. Brody, unpublished.
[c]Coté and Brody 1987a.
[d]Coté and Brody 1987b.
[e]Hubbard and Brody 1975.
[f]P.L. Lakin-Thomas, unpublished.
[g]Juretic 1977.
[h]Luck 1964.
[i]Feldman et al. 1986.
[j]Cote 1986.

2.6.4 Lipids and the Oscillator

What role, if any, do membrane lipids play in the oscillator mechanism? We do not yet have an answer. The data reviewed here enable us to begin consideration of this question. It is clear that defects in membrane lipid metabolism correlate with major defects in rhythmicity in the *Neurospora* mutant strains *cel*, *prd-1* and *chol-1*. Whether other mutants with defects in lipid metabolism also show alterations in rhythmicity is not known, since none have been studied except under conditions of full supplementation, where the lipid defect is not physiologically significant.

The *cel*, *prd-1* and *chol-1* strains all have lengthened periods under conditions where membrane lipid metabolism is defective. In the cases of *cel* and *chol-1* the long periods seen are among the longest "circadian" periods ever reported for any organism. The *cel* and *chol-1* strains, but not the *prd-1* strain, also have similar defects in temperature compensation. The *cel* and *prd-1* strains have abnormal temporal patterns of phospholipid fatty acid composition, while the *chol-1* strain has abnormal phospholipid head group composition. The *prd-1* mutation is epistatic to the *cel* mutation, suggesting that they may affect rhythmicity through alterations in the same pathway. Whether *chol-1* and *cel* or *chol-1* and *prd-1* likewise interact has not been determined.

Period lengthening in the *cel* mutant can be reversed by several factors affecting mitochondrial metabolism, including nonfermentable carbon sources, antimycin A, and the [*mi*], oli^r, and *prd-1* mutations. These mutations and antimycin A also increase levels of mitochondrial protein. These results might implicate mitochondrial membranes in rhythmicity. Alternatively, lipid metabolism in the mitochondria might modulate the effects of the *cel* mutation; this possibility is particularly interesting in light of the evidence for a fatty acid elongation system in mitochondria independent of the cytosolic fatty acid synthase.

On the other hand, there is no simple correlation between the alterations in membrane composition which are seen in *cel*, *prd-1* and *chol-1* strains and the changes in period in these strains. The ability of fatty acids to alter the membrane fatty acid composition of the *cel* strain, for example, does not correlate with their ability to alter the period of this strain; in fact, membrane composition appears to be more normal under conditions where period is most abnormal. Similarly, preliminary evidence indicates that changes in the neutral lipids of *chol-1* better correlate with altered period than changes in the membrane phospholipids. It is possible that changes in bulk membrane composition in these strains are coincidental to changes in rhythmicity, and that other alterations in lipid metabolism — changes in DAG composition or levels, for example — alter the clock.

2.6.5 Lipids and Temperature Compensation

Does temperature compensation of membrane properties contribute to temperature compensation of circadian rhythms? This is an attractive hypothesis, but one which still remains untested. Three different mutants in *Neurospora* appear to have lost temperature compensation (see Fig. 2), *cel*, *chol-1*, and the null alleles of *frq* (frq^9 and frq^{10}). Two of these, *cel* and *chol-1*, have altered membrane lipid composition. Membrane fatty acid composition is altered in *cel*, while phospholipid head group composition is altered in *chol-1*. It seems likely that these strains are defective in the normal variation of membrane composition with temperature, although this has not been tested.

Although all three mutants have lost temperature compensation, the temperature sensitivity of the rhythm differs between *frq* null strains on the one hand, and *cel* and *chol-1* on the other. This suggests that *frq* affects temperature compensation by a different mechanism than the other two mutations. There is, so far, no evidence for altered membrane composition in any *frq* mutant, including the *frq* null mutants, although this has not been completely ruled out either. The partial epistasis between *frq* and *cel* indicates that the two mutations are not completely independent, but each in some way affects the same aspect of metabolism. It is not yet known whether *frq* and *chol-1*, or *cel* and *chol-1* also show partial epistasis in their interactions.

References

Aaronson LR, Martin CE (1983) Temperature-induced modifications of glycosphingolipids in plasma membranes of Neurospora crassa. Biochim Biophys Acta 735:252-258

Aaronson LR, Johnston AM, Martin CE (1982) The effects of temperature acclimation on membrane sterols and phospholipids of Neurospora crassa. Biochim Biophys Acta 713:456-462

Actis Dato SM, Catala A, Brenner RR (1973) Circadian rhythm of fatty acid desaturation in mouse liver. Lipids 8:1-6

Adamich M (1976) Membrane circadian rhythms in Gonyaulax polyedra. Thesis, University of California, Santa Barbara

Aronson BD, Johnson KA, Dunlap JC (1994a) Circadian clock locus frequency: protein encoded by a single open reading frame defines period length and temperature compensation. Proc Natl Acad Sci USA 91:7683-7687

Aronson BD, Johnson KA, Loros JJ, Dunlap JC (1994b) Negative feedback defining a circadian clock: autoregulation of the clock gene frequency. Science 263:1578-1584

Berridge MJ (1987) Inositol triphosphate and diacylglycerol: Two interacting second messengers. Ann Rev Biochem 56:159-193

Berridge MJ, Downes CP, Hanley MR (1989) Neural and developmental actions of lithium: A unifying hypothesis. Cell 59:411-419

Billah MM, Anthes JC (1990) The regulation and cellular functions of phosphatidylcholine hydrolysis. Biochem J 269:281-291

Borgeson CE, Bowman BJ (1985) Blue light-reducible cytochromes in membrane fractions from Neurospora crassa. Plant Physiol 78:433-437

Borgeson CE, Bowman BJ (1990) Mutations that affect circadian rhythms in Neurospora crassa can alter the reduction of cytochromes by blue light. J Biol Rhythms 5:291-301

Brody S, Forman L (1980) Interactions between exogenous unsaturated fatty acids and mitochondria in Neurospora crassa. Abstr Ann Mtg Amer Soc Microbiol 1980:157

Brody S, Martins SA (1979) Circadian rhythms in Neurospora crassa: effects of unsaturated fatty acids. J Bacteriol 137:912-915

Brody S, MacKenzie L, Chuman L (1987) Circadian rhythms in Neurospora crassa: the effects of mitochondrial mutations and inhibitors. Genetics 116:s30

Brody S, Coté GG, Germain J (1989) Circadian rhythms in Neurospora crassa: the effects of increased mitochondrial amounts and metabolism. J Cell Biol 107:348a

Browse J, Roughan PG, Slack CR (1981) Light control of fatty acid synthesis and diurnal fluctuations of fatty acid composition in leaves. Biochem J 196:347-354
Cossins AR, Raynard RS (1987) Adaptive responses of animal cell membranes to temperature. In: Bowler K, Fuller BJ (eds) Temperature and Animal Cells. Company of Biologists, Cambridge, pp 95-111
Coté GG (1986) Circadian rhythms in Neurospora crassa: studies on fatty acid metabolism. Ph.D. Thesis, University of California, San Diego
Coté GG, Brody S (1987a) Circadian rhythms in Neurospora crassa: a clock mutant, prd-1, is altered in membrane fatty acid composition. Biochim Biophys Acta 904:131-139
Coté GG, Brody S (1987b) Circadian rhythms in Neurospora crassa: membrane composition of a mutant defective in temperature compensation. Biochim Biophys Acta 898:23-36
de Gomez Dumm INT, de Alaniz MJT, Brenner RR (1984) Daily variations of the biosynthesis and composition of fatty acids in rats fed complete and fat-free diets. Lipids 19:91-95
Dharmananda (1980) Studies on the Circadian clock of Neurospora crassa: light-induced phase shifting. Thesis, University of California, Santa Cruz
Díaz-Muñoz M, Hernández-Muñoz R, Suárez J, Chagoya de Sánchez V (1985) Day-night cycle of lipid peroxidation in rat cerebral cortex and their relationship to the glutathione cycle and superoxide dismutase activity. Neuroscience 16:859-863
Dunlap JC (1993) Genetic analysis of circadian clocks. Annu Rev Physiol 55:683-728
Dunlap JC, Feldman JF (1988) On the role of protein synthesis in the circadian clock of Neurospora crassa. Proc Natl Acad Sci USA 85:1096-1100
Edmunds LN Jr. (1988) Cellular and Molecular Bases of Biological Clocks. Springer-Verlag, New York.
Elovson J (1975) Purification and properties of the fatty acid synthetase complex from Neurospora crassa, and the nature of the fas$^-$ mutation. J Bacteriol 124:524-533
Engelmann W, Schrempf M (1980) Membrane models for circadian rhythms. Photochem Photobiol Rev 15:49-86
Feldman JF (1982) Genetic approaches to circadian clocks. Ann Rev Plant Physiol 33:583-608
Feldman JF, Pierce SB, Brown D (1986) Temperature compensation in circadian clock mutants of Neurospora. Plant Physiol 80 (suppl):92
Gardner GF (1979) Oscillations of fatty acids in Phaseolus: implications for a membrane theory of biological rhythms. Thesis, Yale University, New Haven
Gardner GF, Feldman JF (1981) Temperature compensation of circadian period length in clock mutants of Neurospora crassa. Plant Physiol 68:1244-1248
Gardner GF, Galston AW (1977) Fatty acids and circadian rhythms in Phaseolus. Plant Physiol 59 (Suppl.):30
Gooch VD, Wehseler RA, Gross CG (1994) Temperature effects on the resetting of the phase of the Neurospora circadian rhythm. J Biol Rhythms 9:83-94
Graber R, Sumida C, Nunez EA (1994) Fatty acids and cell signal transduction. J Lipid Mediat 9:91-116
Harwood JL (1983) Adaptive changes in the lipids of higher-plant membranes. Biochem Soc Trans 11:343-346
Henry SA, Keith AD (1971) Saturated fatty acid requirer of Neurospora crassa. J Bacteriol 106:174-182
Hubbard SC, Brody S (1975) Glycerophospholipid variation in choline and inositol auxotrophs of Neurospora crassa. Internal compensation among zwitterionic and anionic spcies. J Biol Chem 250:7173-7181
Jerebzoff S, Vanden Driessche T (1983) Rhythme circadien de la teneur en acides gras dans la cellule entière et les chloroplastes d'Acetabularia mediterranea. C R Acad Sci 296:319-322

Johnston AM, Aaronson LR, Martin CE (1982) The effects of altered levels of phosphatidylcholine and phosphatidylethanolamine on fatty acid desaturase activity and sterol metabolism during temperature acclimation in a choline auxotroph of Neurospora crassa. Biochim Biophys Acta 713:512-518

Juretic D (1977) The effect of phosphatidylcholine depletion on biochemical and physical properties of a Neurospora crassa membrane mutant. Biochim Biophys Acta 469:137-150

Klemm E, Ninnemann H (1978) Correlation between absorbance changes and a physiological response induced by blue light in Neurospora. Photochem Photobiol 28:227-230

Lakin-Thomas PL (1992) Phase resetting of the Neurospora crassa circadian oscillator: effects of inositol depletion on sensitivity to light. J Biol Rhythms 7:227-239

Lakin-Thomas PL (1993a) Effects of inositol starvation on the levels of inositol phosphates and inositol lipids in Neurospora crassa. Biochem J 292:805-811

Lakin-Thomas PL (1993b) Evidence against a direct role for inositol phosphate metabolism in the circadian oscillator and the blue-light signal transduction pathway in Neurospora crassa. Biochem J 292:813-818

Lakin-Thomas PL (1995) Disruption of the Neurospora crassa circadian oscillator by choline depletion. Submitted

Lakin-Thomas PL, Brody S (1985a) Circadian rhythms in Neurospora crassa: interactions between clock mutations. Genetics 109:49-66

Lakin-Thomas PL, Brody S (1985b) A pantothenate derivative is covalently bound to mitochondrial proteins in Neurospora crassa. Eur J Biochem 146:141-147

Lakin-Thomas PL, Coté GG, Brody S (1990) Circadian rhythms in Neurospora crassa: biochemistry and genetics. Crit Rev Microbiol 17:365-416

Lakin-Thomas PL, Brody S, Coté GG (1991) Amplitude model for the effects of mutations and temperature on period and phase resetting of the Neurospora circadian oscillator. J Biol Rhythms 6:281-297

Lecharny A, Tremolières A, Wagner E (1990) Correlation between the endogenous circadian rhythmicity in growth rate and fluctuations in oleic acid content in expanding stems of Chenopodium rubrum L. Planta 182:211-215

Liscovitch M (1992) Crosstalk among multiple signal-activated phospholipases. Trends Biochem Sci 17:393-399

Liscovitch M, Cantley LC (1994) Lipid second messengers. Cell 77:329-334

Loros JJ, Feldman JF (1986) Loss of temperature compensation of circadian period length in the frq-9 mutant of Neurospora crassa. J Biol Rhythms 1:187-198

Luck DJL (1964) The influence of precursor pool size on mitochondrial composition in Neurospora crassa. J Cell Biol 24:445-460

Lynch DV, Thompson GA Jr. (1984) Retailored lipid molecular species: a tactical mechanism for modulating membrane properties. Trends Biochem Sci 9:442-445

Lyudnikova TA, Chernysheva EK, Bezzubov AA, Kritsky MS (1989) Possible relationship of cryptochrome photoreceptor to the fatty acid desaturation complex. Orig Life Evol Biosph 19:422

Maresca B, Cossins AR (1993) Fatty feedback and fluidity. Nature 365:606-607

Martin CE, Johnston AM (1983) Changes in fatty acid distribution and thermotropic properties of phospholipids following phosphatidylcholine depletion in a choline-requiring mutant of Neurospora crassa. Biochim Biophys Acta 730:10-16

Martin CE, Siegel D, Aaronson LR (1981) Effects of temperature acclimation on Neurospora phospholipids. Fatty acid desaturation appears to be a key element in modifying phospholipid fluid properties. Biochim Biophys Acta 665:399-407

Mattern DL (1985a) Preparation of functional group analogs of unsaturated fatty acids and their effects on the circadian rhythm of a fatty-acid-deficient mutant of Neurospora crassa. Chem Phys Lipids 37:297-306

Mattern DL (1985b) Unsaturated fatty acid isomers: effects on the circadian rhythm of a fatty-acid-deficient Neurospora crassa mutant. Arch Biochem Biophys 237:402-407
Mattern D, Brody S (1979) Circadian rhythms in Neurospora crassa: effects of saturated fatty acids. J Bacteriol 139:977-983
Mattern DL, Forman LR, Brody S (1982) Circadian rhythms in Neurospora crassa: a mutation affecting temperature compensation. Proc Natl Acad Sci USA 79:825-829
Mikolajczyk S, Brody S (1990) De novo fatty acid synthesis mediated by ACP in Neurospora crassa mitochondria. Eur J Biochem 187:431-437
Murphy DJ, Stumpf PK (1980) Polyunsaturated fatty acid biosynthesis in cotyledons from germinating and developing Cucumis sativus L. seedlings. Plant Physiol 66:660-665
Nakashima H (1983) Is the fatty acid composition of phospholipids important for the function of the circadian clock in Neurospora crassa? Plant Cell Physiol 24:1121-1127
Nakashima H (1987) Comparison of phase shifting by temperature of wild type Neurospora crassa and the clock mutant, frq-7. J Interdiscipl Cycle Res 18:1-8
Njus D, Sulzman FM, Hastings JW (1974) Membrane model for the circadian clock. Nature 248:116-120
Oda K, Hasunuma K (1994) Light signals are transduced to the phosphorylation of 15 kDa proteins in Neurospora crassa. FEBS Lett 345:162-166
Prior SL, Robson GD, Trinci APJ (1994) Phosphoinositide turnover does not mediate the effects of light or choline, or the relief of derepression of glucose metabolism in filamentous fungi. Mycol Res 98:291-294
Rensing L, Bos A, Kroeger J, Cornelius G (1987) Possible link between circadian rhythm and heat shock response in Neurospora crassa. Chronobiol Intl 4:543-549
Rensing L, Kohler W, Gebauer G, Kallies A (1993) Protein phosphorylation and circadian rhythms. In: Battey NH, Dickinson HG, Hetherington AM (eds) Post-translational Modifications in Plants. Cambridge University Press, Cambridge, pp 171-185
Rikin A, Dillwith JW, Bergman DK (1993) Correlation between the circadian rhythm of resistance to extreme temperatures and changes in fatty acid composition in cotton seedlings. Plant Physiol 101:31-36
Roeder PE, Sargent ML, Brody S (1982) Circadian rhythms in Neurospora crassa: oscillations in fatty acids. Biochemistry 21:4909-4916
Scarborough GA, Nyc JF (1967) Methylation of ethanolamine phosphatides by microsomes from normal and mutant strains of Neurospora crassa. J Biol Chem 242:238-242
Schlame M, Brody S, Hostetler KY (1993) Mitochondrial cardiolipin in diverse eukaryotes. Comparison of biosynthetic reactions and molecular acyl species. Eur J Biochem 212:727-735
Scholübbers HG, Taylor W, Rensing L (1984) Are membrane properties essential for the circadian rhythm of Gonyaulax? Am J Physiol 247:R250-R256
Thompson GA Jr. (1983) Mechanisms of homeoviscous adaptation in membranes. In: Cossins AR, Sheterline P (eds) Cellular Acclimatisation to Environmental Change. Cambridge University, Cambridge, pp 33-53 (Soc Exp Biol Symp, Seminar Ser, vol 17)
Vigh L, Los DA, Horváth I, Murata N (1993) The primary signal in the biological perception of temperature: Pd-catalyzed hydrogenation of membrane lipids stimulated the expression of the desA gene in Synechocystis PCC6803. Proc Natl Acad Sci USA 90:9090-9094
Vokt JP, Brody S (1985) The kinetics of changes in the fatty acid composition of Neurospora crassa lipids after a temperature increase. Biochim Biophys Acta 835:176-182

3 Membrane Glycosylation and Circadian Rhythms in Plant Systems and in Animal Normal and Transformed Systems

Ghislaine M. Petiau-de Vries
Faculté de Médecine, Chimie Générale I, Université Libre de Bruxelles, Bruxelles, Belgium[1]

Summary

A review is given on the properties of glycoconjugates subjected to circadian rhythms: proteoglycans, glycosylated hormones, enzymes, glycolipids. Membrane receptors and their ligands which appear to be frequently glycosylated can play a role in circadian rhythmicity. Circadian modulation of the glycoconjugates metabolism and the effect of glycosylation inhibitors on the circadian rhythms will be discussed. Rhodopsin, a photoreceptor, is glycosylated and presents an unusual type of glycosylation. The role of ceramides as second messengers has been demonstrated and could interact with metabolites important for the rhythm. The formation of a spatial structure of macromolecules such as helicoids is a circadian or ultradian process. The importance of circadian rhythms in cancer has been reported in a few cases and suggests a critical analysis of many reports.

---

[1] Abbreviations : Ach acetylcholine; AChE acetylcholinesterase; CEA carcinoembryonic antigen; cAMP cyclic adenosine 5'-monophosphate; CREM cAMP -responsive element modulator; ELISA enzyme-linked immunosorbent assay; EPO erythropoietin; FSH follicle stimulating hormone; gal galactose; galNAc N-acetylgalactosamine; GH growth hormone; Glc glucose; GM1 Gal beta 1- 3 GalNAc beta 1-4 Gal (NeuAc alpha 2-3) beta 1-4 Glc beta 1'-ceramide; cGMP cyclic guanosine 5'-monophosphate; 5-HT 5-hydroxytryptamine or serotonin; LH luteinizing hormone; man mannose; NeuAc sialic acid; PRL prolactin; TG threonine/glycine.

3.1 Introduction

Circadian rhythms in Eukaryotes appear to modulate cell division, enzyme activities, body temperature, activity, mitotic index, ion uptake, sensitivity to light or temperature, sensitivity to metabolic inhibitors, etc. Different models have been proposed that were grouped into four main categories by Edmunds (1988; 1992). In the network models, the rhythm could result from control loop hypothesis. They may include molecules exhibiting different structures: for example a phosphorylated and a dephosphorylated form. In the transcriptional models such as the chronon model, the circadian period should result from rate-limiting diffusion steps by mRNA to the ribosomes. In the category referred to as the membrane models, ions or protons gradients, passive diffusion of hypothetical X molecule or pumps play a significant role.

Schweiger and Schweiger (1977) included protein synthesis in the coupled translation-membrane model.

Alternative models have been developped. A recent one proposed that circadian rhythms derive from controlled chaotic cellular oscillators (Klevecz and Bolen 1993 ; Lloyd and Lloyd 1993). An other view is based on multiple cellular oscillators. A recent one was developped by Vanden Driessche (1994) in which it is assumed that the expression of the clock depends on multiple cellular oscillators coupled with a complex network of synchronizing pathways (see also Introduction and overview, this volume, for references to very recent models).

In 1982, circadian rhythm of enzymes participating to the synthesis of oligosaccharides in glycoproteins was reported (Louisot 1982) .

Galoutzov (1985) showed that glycoconjugates were involved in circadian rhythms. This has been confirmed by Vanden Driessche et al. (1992) who observed altered rhythmicity after treatment of Acetabularia with glycosylation inhibitors.

A number of glycoconjugates presenting circadian rhythmicity have been observed (Louisot, 1982; Wide et al.,1989, Focan, 1987).

Thus, several questions are put forward.

First, are these compounds embedded in the plasma membrane where they function as receptors or are they transiently associated with a membrane receptor -what is their exact role-?

Secondly, the glycoconjugates presenting an extraordinary source of diversity and potential for power regulation, is this diversity important for the clock? In the examples presented here, if we assume that the rhythmic response is composed of an input, of the synchronizing signal transduction pathways, of one or several pacemakers and output pathways, we can observe that glycoconjugates are present at various steps. For example, light, the most studied synchronizing agent , interacts with photoreceptors such as rhodopsin, which is a glycoprotein. Several hormones or plasma membrane receptors displaying circadian rhythmicity are glycoproteins. Glycoconjugates, some of which are under circadian control, participate in the transmission of information from the external milieu to the cytoplasm, from cell to cell or from cells to the extracellular matrix.

The transcription of a number of clock genes is under circadian control in higher plants and animals (reviewed by Dunlap 1993 and Takahashi 1994). The cDNA sequences of the *per* gene reveal that the gene products contain putative N- and O-glycosylation sites and threonine/glycine (TG) repeats. The TG repeats are found in proteoglycans.

3.2 Glycoproteins

3.2.1 Glycosylation

The glycoproteins are ubiquitous biopolymers constituted by polypeptides containining one or several oligosaccharidic moieties covalently linked by a complex enzymatic machinery. Plant and animal oligosaccharides attached to the glycoproteins display homologies. Both are characterized by a extraordinary diversity resulting from the existence of different monosaccharides (about ten) linked by different types of glycosidic linkages (at least five).

In plants and animals, the oligosaccharides differ by their length and by the "branching" , being linear or not. If we compare the diversity resulting from the formation of a small peptide composed of three amino acid to that obtained for three sugars, we can form six different peptides whereas 1056 possibilities exist for the oligosaccharides. Owing to the technological progresses of the last decade, thousands of structures have been described. However, glycobiologists are unable to explain why different oligosaccharidic chains can be linked to the same polypeptide.

Two major classes of structures are present sometimes on the same polypeptide: the N-linked oligosaccharides where the first sugar, a N-acetylglucosamine is linked to an asparagine, in plant and animals and the O-linked oligosaccharides where the first sugar (usually a N-acetylgalactosamine or a N-acetylglucosamine) is linked to a threonine or a serine in animal or to a hydroxyproline or a serine in plants (reviewed by Driouich et al 1993). The synthesis of the N-linked oligosaccharides begins in the endoplasmic reticulum and is achieved in the Golgi apparatus. O-linked moieties are synthetized exclusively in the Golgi.

The synthesis is a complex process including a lipidic intermediary in the case of the N-linked oligosaccharides and one specific enzyme per glycosidic link formed. These N-linked sugars are only found when the asparagine residues is part of a sequence: asparagine-X-serine/threonine, a glycosylable site.

Carbohydrates confer to glycoproteins interesting properties such as resistance against proteolytic degradation, stability, water-binding, recognition (cell-microbes, cell-cell, cell-matrix) and antigenic properties. This list is not exhaustive and many roles ascribed to the presence of carbohydrate portions are probably correct, paraphrasing Varki (1993): Biological roles of oligosaccharides: all of the theories are correct.

Analyzing the literature on glycoconjugates, it becomes obvious that no unitary theory can be deduced with simple rules, by which structure and the exact role of specific oligosaccharides can be predicted and regulatory mechanisms of their synthesis be understood. An unsolved problem is the functional significance of the numerous variants existing for a single polypeptide chain.

3.2.2 Glycosylation and Circadian Rhythms

The possible importance of glycoconjugates in circadian rhythms in plants was described by Galoutzov (1985) who observed the perturbation of Acetabularia development in the presence of inhibitors of glycosylation and by Vanden Driessche et al. (1992). They carried out a study of effects of inhibitors of glycosylation on growth and morphology taking time into account. They also reported that the photosynthesis circadian rhythm in Acetabularia was affected by tunicamycin in a time-dependent way.

Many receptors are glycoproteins. They are present in animal plasma membranes. Among them, hormone receptors represent an important class which will be considered in section 3.2.5, devoted to hormones. Directly involved in the circadian response, the light receptors present in plants and animals represent a class of proteins frequently bound to membranes. Their interaction with light, one of the best studied environmental signal, is the primary biochemical interaction which is important for the biological clock. However, the exact role of the glycosyl moieties is unknown.

3.2.3 Light Receptors

Plants. *Phytochromes.* Circadian rhythms in plant are entrained by diurnal light-dark cycle involving the participation of photoreceptors such as the phytochromes in certain systems (reviewed by Lumsden, 1991), chromoproteins composed of a dimer of 120-kD subunits. They contain a covalently attached tetrapyrrole chromophore. Soluble forms are described which can be transiently attached to membranes. For the times being, however, binding has to be considered with caution (Chory, 1994). Interestingly, early steps in red light transduction involve activation of a G protein (Neuhaus et al.,1993). DNA blot analysis of phytochromes show the existence of several gene sequences, some of them sharing N-glycosylable sites (Sharrock and Quail 1989).

Animals. *Rhodopsin.* Phototransduction in the vertebrate retina is dependent on cyclic GMP-activated ionic channels in the plasma membrane of rods and cones. Exposure to light entrains a circadian oscillator (Kaupp et al. 1988). The first step of light transduction is known to imply photosensitive receptors.

Rhodopsin, the light receptor in retinal rod cells is the most extensively characterized member of a superfamily of G-protein coupled receptors.

Gt protein (or transducin) acts as an intermediary between rhodopsin and activates the effector protein cGMP phosphodiesterase.The subsequent lowering of cGMP closes the cation-specific channels and hyperpolarize the membrane (Bourne et al. 1990). Rhodopsin consists of a protein with several glycosylations sites: opsin (Mr 39kDa). A photosensitive 11-cis retinaldehyde prosthetic group is covalently attached to the molecule. It represents 90% of the disk membrane of the rod outer segments. A rhodopsin-like protein named pinopsin is also present in the chicken pineal gland (Okano et al. 1994).

It is interesting to compare the polypeptides and the oligosaccharides of the rhodopsins which are highly conserved from Drosophila to mammals as demonstrated by the cDNA analysis (Korenbrot and Fernald 1989; Hargrave and McDowell,1992).

Two N-linked oligosaccharides are present in frog (Rana pipiens) rhodopsin, 60 % being constituted by: Glc NAc alpha 1-2 Man beta 1-3 (Man beta1-6) Man alpha1-4 GlcNAc alpha1-4 GlcNAc-Asn. The remaining structures contain hexoses and some include sialic acid which was absent in the bovine protein (Duffin et al.1993). Fujita et al. (1994) also reported the presence of sialic acid in human rhodopsin. It may be important to note that rhodopsin plays a role in adhesion during morphogenesis. Smith et al.(1990) reported a transient galactose fixation which may be essential for rod outer segment morphogenesis. The galactose, one of the ultimate or penultimate sugar residues, is attached to the oligosaccharide by a galactosyltransferase. Ju and Kean (1992) reported that, *in vitro*, the galactose transfer to rhodopsin was low in the presence of bovine retina galactosyltransferase in contrast with the milk galactosyltransferase.

A surprising observation made by these authors was that after bleaching by visible light, opsin was preferred over rhodopsin as an acceptor of galactose by the galactosyltransferase of bovine and embryonic chick retina and by the enzyme of rat liver. This distinction was not observed for the milk enzyme. It is not sufficient to deduce that galactose fixation plays a regulatory role but suggests that carbohydrate structure could play some biological role in opsin.

Moreover, the chick embryonic galactosyltransferase transfers galactose to an unusual acceptor: an oligosaccharide from rhodopsin containing five mannose residues.

During the formation of N-linked oligosaccharides, a structure containing nine mannose residues is transiently present but only three mannose residues are observed in the oligosaccharides that accept additional sugars. The authors suggest that this novel oligosaccharide could be specific to photoreceptors. This offers interesting possibilities for chronobiologists and biochemists.

3.2.4 Serotonin Receptor

The neurotransmitter, serotonin or 5-hydroxytryptamine (5-HT) is released by nerve terminals in central and peripheral nervous system. Its binding to cell-surface receptors activates second messsenger pathways. Several serotonin receptors have been defined according to their pharmacological properties. One of these receptors, the 5-HT_{1c} from *Xenopus* oocytes liberates inositolphosphate that raises Ca^{2+} level, leading to the opening of Ca^{2+} dependent chloride channels (reviewed by Julius et al 1988). The receptor possesses one N-linked glycosylable site. Serotonin is the precursor of melatonine, the main hormone secreted in the pineal gland in response to dark/light transition. It also plays a role in phase shifting the rhythm in the *Aplysia* eye, a model system as well as in a number of systems (see for example Arendt 1992) . The 5-HT entrainment pathway involves effect on protein synthesis, increase in cAMP, activation of adenylate cyclase and hyperpolarization (Koumenis and Eskin 1992). The predicted structure of the Drosophila serotonin receptor (5-HT-drosophila receptor) reveals eight transmembrane domains, two possible N-glycosylation sites and a Gly-Ser repeat that is a potential glycosaminoglycan attachment site. After introduction into mice NIH 3T3, the gene activates adenylate cyclase in response to serotonin (Witz et al.(1990). The role of serotonin-sensitive cyclase in the circadian rhythmicity of invertebrate systems has been suggested by Kandel et al., 1983 and reviewed by Takahashi and Zatz 1982. The cDNA sequence of the first putative extracellular domain of the 5-HT-drosophila receptor exhibits homology with a rat yolk sac tumor chondroitin sulfate proteoglycan (Bourdon et al.,1985), with a Drosophila clock gene, period (per) (Bargiello and Young 1984; Zehring et al.1984) and with a Neurospora clock gene (frq) (McClung et al. 1989).

A physiological importance of the glycans in this receptor formation is an attractive possibility.

3.2.5 Hormones and Receptors

Plants. Hormones bind to receptors in order to achieve activity. Little is known about these receptors and they are accordingly briefly examined by Vanden Driessche et al. (this volume). The ethylene receptor is presumably not glycosylated and no information whatsoever is available for the other receptors .

Animals. Subject to intense research for years, an enormous amount of information has accumulated and the glycosylation process appears to be promising in the understanding of the way in which they modify hormone activity, possibly, at least in some cases, implementing CR on these activities.

Endocrine glycoproteic hormones are synthesized in specialized glands, released into body fluids and carried to their target tissue. As the first step in their regulatory role, they associate with specific receptors on the plasma membrane.

Growth Hormone. Circadian variation of the pituitary growth hormone plasma concentration in man has been demonstrated by Drobny et al.(1983). Hepatocytic plasma membranes possess two types of receptors: the somatogenic binding sites specific for the growth hormone (GH) and the lactogenic binding sites recognized by GH and prolactin, the hormone stimulating milk secretion in lactating glands. Maximal occupancy of somatogenic rat liver binding sites was reached at the time of peak serum GH. The circadian rhythm in binding activity results from alteration of the number of binding sites, the affinity being constant. The study was performed during 5 hours, between 9:00 - 14:00. Thus, the centrally regulated GH secretion is further controlled by the dynamics of the GH receptors turnover (Bick et al. 1989). Mode et al.(1982) suggested that the temporal pattern of GH secretion was the major determinant in the sexual differentiation of hepatic steroid metabolism.

Lectin inhibition of GH binding was observed in the liver of female mice, suggesting the involvement of glycosylated structures at the receptor sites (Lascols et al. 1983).

In the liver, the GH controls the production of the insulin-like growth factor-I which is produced as a prohormone and contains one glycosylation site (Dull et al.1984) . Insulin and insulin-like growth factor I receptors are also glycoproteins and total deglycosylation suppresses the binding activity. Partial change in affinity or specificity occurs with partial deglycosylation (Duronio et al.1986; 1988). Thus glycosylation appears to play a key role in receptor activity.

The carbohydrate structures may be important in GH secretion, in GH-GH receptor interactions as well as in GH receptor turnover .

Erythropoietin (EPO) is a haematopoietic serum growth hormone, synthesized and secreted by the kidney in response to anaemia and hypoxia. Altough some studies demonstrate circadian variations in serum EPO in normal subjects (Wide et al. 1989; Klausen et al. 1993), other authors have failed to observe diurnal variations (Miller et al. 1981). The circadian rhythm has been observed in patients with various diagnosis excluding polycytaemia and kidney diseases (Wide et al. 1989) and in patients with hypoxaemic sleep apnoea (Mc Keon et al. (1990) where the am and the pm levels differed from those in patients with lung disease (Miller et al. 1981). Altough there is a discrepancy between the results, we cannot conclude that some of them are more pertinent than others. Comparison of the assays show important differences. Several authors mention the origin of the EPO used to prepare the antibody: Wide et al.(1989) used a polyclonal antibody raised in a rabbit with human urinary EPO. In a radioimmunoassay, the human recombinant EPO was radioiodinated. In the radioimmunoassay used by Miller et al., pure human EPO obtained from the urine of severely anaemic patients was iodinated. In the two-sites ELISA commercial tests (McKeon et al. 1990, Klausen et al. 1993), the antibodies were probably monoclonal.

Discrepancies can thus be attributed to heterogeneity of EPO due to the existence of isoforms, differing with respect to the nature of the oligosaccharides. Up to 30 different isoforms (Takeuchi and Kobata 1991) can be obtained in recombinant EPO. At present, no specific isoform or group of isoforms can be assigned to EPO from unhealthy or healthy subjects. The results obtained in each group are probably correct for any type of assay. Takeuchi and Kobata (1991) proposed the following hypotheses with regard to the functional role of carbohydrate chains in EPO. According to their theory, active hormones should contain lower quantities of sialic acid. EPO with a high sialic acid content may form a pool of precursor which could be activated by a gradual removal of sialic acid in a time-dependent manner.

The EPO receptor may be specific of some EPO oligosaccharides. We suggest that the changes of the EPO glycopeptides with time should be studied in order to test the hypothesis of a circadian regulated glycosylation of this hormone.

Luteinizing Hormone and Follicle-Stimulating Hormone. These two hormones are secreted in the anterior pituitary and transported by circulation to the ovary.

These proteins are members of the glycoprotein hormone family constituted by heterodimers sharing a common alpha subunit.

Pulsatile and circadian pattern of luteinizing hormone (lutrophin, LH), follicle-stimulating hormone (FSH) and testosterone release have been identified in normal men (Veldhuis 1987); LH and FSH increases coincided.

LH. The existence of hormone variants whith different isoelectric points, suggests differences in the sialic acid content (Ujihara et al. 1992). In women, the extent of glycosylation has been shown to vary according to the menstrual cycle (Padmanabhan et al. 1988) but the various isoforms at different times of day have not been analyzed. Such a study might provide valuable information in regard to the role of glycosylation in biological rhythms. Some porcine LH isoforms could stimulate kidney growth suggesting a functional role for oligosaccharides (Nomura et al. 1988).

The nonspecific binding of LH and other glycoprotein hormones to gangliosides (see section 3.3) was reported by Karlsson (1977). This suggests that these gangliosides might be cofactors of a specific polypeptide receptor.

FSH. It should be noted that FSH not only functions as a hormone but that it also controls the accumulation of CREM (cyclic AMP-responsive element modulator) messenger RNA: the CREM function is converted from an antagonist to an activator; this is due to a switch in the use of polyadenylation sites leading to a greater stability of the transcript (Foulkes et al., 1993; Stehle et al. 1993). Owing to the importance of CREM genes which encode a family of activators and repressors of cAMP transcription and play a key role in the hypothalamic-pituitary-gonadal axis, the hormonal control mechanism should be studied more deeply. Moreover, FSH glycosylated variants should be actively studied in order to test the hypothesis of different roles for the proteins carrying different oligosaccharides.

Prolactin. Meier (1973) reported that injections of prolactin (PRL) and corticosterone, differently timed, mimicked the effect of long and short days.

Two human prolactin structural variants have been described in human serum : a glycosylated (G-PRL) and a non-glycosylated form. The hormone levels in the morning in a fasting state and in the afternoon in a non-fasting state did not significantly differ but this does not allow a clear-cut conclusion (Warner et al. 1993). The properties of the prolactin receptor, the lactogenic site of GH were discussed in this section.

3.2.6 Enzymes

Plants. *Acetylcholinesterase.* A number of plants in a variety of families contain acetylcholine altough its function remains hypothetical as well as its mode of action. Moreover it seems possible, on the basis of their insensitivity to Ach inhibition, that a number of other plants do not contain ACh. Cholinesterase was detected in several plants but a characterization was only described by Riov and Jaffe (1973) in mung bean roots. It has been shown (Tretyn and Tretyn 1990) that acetylcholine (ACh) is present in etiolated and green oat (Avena sativa L. cv Diadem) seedlings.The changes in ACh level were correlated to that of acetylcholinesterase (AChE). In the etiolated coleoptiles, the products of enzymatic reaction showing AChE activity accumulated on the external side of plasma membrane. This activity was regulated by light and changes of the green plants were circadian under a LD 16:8 cycle. The presence of carbohydrate epitopes in the enzyme may be important for its adhesive properties (Schachner 1989).

Animals. *Acetylcholinesterase.* In mice, a circadian rhythmicity of AChE has been demonstrated in the brain stem reticular formation under normal and constant regimens (Lewandowski 1988). The receptor for acetylcholine has been shown to display a circadian rhythm in brain. Moreover, this rhythmicity is correlated with that of glycosyltransferases, the enzymes responsible for protein glycosylation (Broquet et al. 1987). it would be interesting to examine whether the correlation corresponds to a causal relationship.

3.2.7 Extracellular Matrix

Plants and Animals. The extracellular matrix (ECM) is constituted mainly by polymers that surround cells and provide protection, support or a storage area. Moreover, the ECM interacts with the membranes in both plant (Roberts 1989) and animal (Bernfield 1989) cells. In animal cells, it is lying under epithelia and surrounds connective-tissue cells. Major proteic components of the ECM are collagens, non-collagenous glycoproteins, proteoglycans, elastins , cell-adhesion molecules.

The interactions between plasma membrane and ECM could play a crucial role in growth, differentiation, invasiveness. In plants, the ECM represents the cell wall plasma membrane limiting a subcellular compartment, it is composed of polymers such as cellulose or hemicellulose. In spite of these differences, the ECMs from both plant and animal cells share interesting common properties related to the biological rhythms.

Helicoids. Chronobiologists have observed a rather surprising phenomenon : ECM ultrastructures are characterized by common three-dimensional helicoidal arrangements resulting from periodic activities. The oscillatory processes observed may be ultradian, but an adjustment of this endogenous rhythm by external factors having a circadian periodicity could mask the ultradian character. The spatial and temporal regulation in helicoidal ECM has been reviewed by Satiat-Jeunemaitre (1992). In plants, the helicoids are present in epidermis, wood vessels, seed mucilage, spores. In animals, they are found in antennas, procuticle, basement membrane, stroma of cornea. Their components are attached to preexisting structures and grow with time. They are composed of two phases: a rigid phase with low elasticity such as cellulose, collagen (which could be glycosylated), keratin-like protein, chitin and a surrounding amorphous phase with low elasticity: hemi-cellulose, proteoglycans, xyloglucans.

All possible combinations for the two phases exist : protein / protein, polysaccharide / polysaccharide, protein / polysaccharide. The periodicity of the rhythm corresponds to a complete rotation of microfibrils that corresponds to the pitch of the helicoid.

Animals. *Interphotoreceptor Matrix.* Within the framework of studies on the transfer of substances between photoreceptors and retinal pigment of rat eyes, Uehara et al.(1990) described a shift in the distribution or molecular conformation of the interphotoreceptor matrix (EPM) after light and dark transition. This was not a circadian effect. The presence of glycoconjugates was demonstrated by the fixation of lectin probes specific for sialic acid or N-acetylglucosamine residues.

3.3 Glycolipids

Glycolipids are ubiquitous membrane components of both plant and animal cells, often located in the outer leaflet of the plasma membrane .

They may function as receptors and play a role in cell-cell recognition, cell transformation and tumor progression but can also display some regulatory role on enzymes or immunologic processes (reviewed by Kundu 1992). Among the natural glycolipids, we find the glycosphingolipids present in plant and animal cells; they are made of a hydrophobic ceramide moiety (N-acylsphingosine) and one or more sugars linked glycosidically to the terminal primary hydroxyl group of sphingosine. Mono- and diacylglycosylglycerol constitute an other class of glycolipids present in plants.

Plants Plants glycolipids are constituted mainly by monoglycosyldiacylglycerol and diglycosyldiacylglycerol as reviewed by Curatolo (1987). If they constitute the major components of thylakoid membranes, small amounts are also found in plasma membranes. Within chloroplasts, thylakoids constitute flattened sacs containing light-energizable complexes. Glycolipids are probably important for the lamellar structure of the thylakoids and the stabilization of the photosynthetic pigment-protein complexes. The photosynthesis displays a circadian rhythmicity in protoctista (see Vanden Driessche this volume). In opposition to mammalian systems where glycolipids appear to be present only in the extracellular membrane monolayer, plant glycolipids have been observed in both layers of the thylakoid bilayer.

Monogalactosyldiacylglycerol could be involved in the fitting the large protein-pigment complexes into the thylakoid membrane.

There are probably small amounts of glycosphingolipids in plant plasma membranes where the sugar moiety is galactose or glucose.

Animals Glycosphingolipids are ubiquitous components of animal cells from bacteria to vertebrates where they are particularly abundant in the nervous system.

They play a role in diverse functions: maintenance of the membrane structure, differentiation, neoplastic transformation and they have been recognized as important modulators for transmembrane signaling (reviewed by Hakomori 1981;1990).

Gangliosides defined as sialic acid containing glycosphingolipids are mainly components of the plasma membrane. At the cell surface, they may interact with a variety of ligands including hormones or influence adenylate cyclase activity. Several gangliosides act as modulators of various kinases (Igarashi et al. 1989). The plasma membranes could modulate the degree of ganglioside sialosylation which in turn could influence cell performance (reviewed by Riboni and Tettamanti 1993).

Gangliosides undergo endocytosis from the plasma membrane to the Golgi apparatus, to the endosomes and to the lysosomes or can return directly to the plasma membrane (this is called retro-endocytosis). Degradation leads to the apparition of sphingosine or ceramide considered now as second messengers, regulating for example protein phosphorylation by exerting a powerful effect on protein kinase C which controls cellular metabolism (Hannun and Bell 1989; Hakomori 1990; Younes et al.1992; Okazaki et al.1994). Owing to these properties, such a type of second messengers should retain the attention of chronobiologists.

3.4 Proteoglycans

Animals Proteoglycans (PGs) and their constituent carbohydrate moieties, glycosaminoglycans (GAGs) are present in all animal tissues. They are important components of the extracellular matrix but are also present in the plasma membrane, the cytoplasm and the nucleus of cells. The proteoglycans appear to be tissue-specific rather than species-specific. Both N- and O-linked oligosaccharides moieties, chondroitin sulfate and heparan sulfate can be present on the core protein (see reviews by Ruoslahti, 1988; Bhavanadan and Davidson, 1992).

It is becoming clear that interactions between proteoglycans and cytokines play an important regulatory role. The relationship between growth factors or cytokines and proteoglycans have not been studied in terms of biological rhythms. The spatio-temporal arrangement of matrix constituents including proteoglycans has been investigated. The specificity of cytokine signalling should result from the interactions between cell surface receptors and with proteoglycans present on the cell surface and on the extracellular matrix (reviewed by Yamaguchi 1993). In Drosophila, circadian activity is controlled in part by the *per* gene coding for a proteoglycan. Young and his collaborators

(Baylies et al., 1993) discussed the possible but questioned significance of the protein.

3.5 Oligosaccharins

Plants In plants some oligosaccharides composed of a few saccharides such as oligo ß-glucosides or oligo ß-galacturonides (Darvill, 1984; Ryan, 1987; Darvill et al.,1992) intervene in plant defence, development or growth and are referred to as oligosaccharins.

If we consider their saccharidic structure, they do not belong « stricto sensu » to the glycoconjugates. However, they represent a relatively new area of research that interests many glycobiologists.

They were detected in culture and in the mycelial walls of plant fungal pathogens. For example, natural and synthetic oligoglucosides can bind to membranes from the major organs of soybean plants. In some cases, this binding is followed by the synthesis of phytoalexins, plant defence molecules and thus act as elicitors (an elicitor is a molecule that induces defence responses in plants) .

The oligosaccharins derived from chitin and chitosan , the components of the cell walls of many fungi and of the exoskeletons of invertebrates. They can elicit defence mechanisms. Oligogalacturonides prepared from pectic polysaccharide and oligo-ß-glucoside act synergistically to elicit phytoalexins. The responses to oligogalacturonides at plasma membrane level is extremely rapid.

These studies are promising in order to develop new strategies for improving plant resistance. Research should concern the natural mechanisms including the participation of oligosaccharins and the isolation or chemical synthesis of new oligosaccharins. The search of rhythmic processes depending on elicitors is certainly important in order to develop efficient techniques of plant defences taking these into account.

3.6 Calcium

Calcium is important for biological rhythms and intervenes at different levels of the rhythmic response, as part of the input or the output pathway or in the clock itself (Edmunds and Tamponnet in a review ,1990).

Ca^{2+} enters into the cell via channels which are activated by the effect of light on photoreceptors. In the cytoplasm, Ca^{2+} exist in free and in sequestered form. In addition, Ca^{2+}, cAMP and inositolphosphate function as second messengers and they may initiate phosphorylation.

Plants Ca^{2+} may form a complex with calmodulin. The hypothesis of circadian-time dependence of the complex Ca^{2+}-calmodulin properties was suggested by the observation of growth modification in algae in the presence of trifluoroperazin, a calmodulin inhibitor (Marmé and Dieter (1983); Cormier et al. (1983) ; Vanden Driessche 1986). The Ca^{2+}-Calmodulin complex influences various kinases and the question of the pleiotropic effect of the second messenger cannot be evaded as well as that of the kinase substrates in connection with rhythmicity.

The role of glycoconjugates has not been demonstrated.

Animals Modulation of a Ca^{2+} signalling pathway by GM1 ganglioside (a glycosphingo- lipid containing sialic acid) has been described in neural cells (Hilbush and Levine 1992). The ganglioside stimulates a Ca^{2+} - dependent protein kinase and enhances depolarization-induced entry of extracellular Ca^{2+}.

The distribution of free ionized calcium between the extracellular and the intracellular compartments in the heart is controlled by the sarcolemma which is abundantly glycosylated and contains sialic acid residues. Neuraminidase treatment which removes the sarcolemmal surface sialic acid residues was found to increase cellular Ca^{2+}. The authors suggested that sialic acid may play a significant role in the regulation of cellular Ca^{2+} and contractile activity (Yee et al. 1991). In an intriguing report, Purushotam et al. (1992) demonstrated that galactosyltransferase was the preferred substrate of a Ca^{2+}-Calmodulin-dependent serine kinase from rat parotid gland acinar cells (after treatment with isoproterenol, a beta agonist). Besides its enzymatic function in the Golgi apparatus, galactosyltransferase has a plasma membrane form which may interact with laminin from basement membranes (Runyan et al. 1988).

These observations suggest that glycosylation is important for Ca^{2+} metabolism. It should be interesting to inhibit glycosylation and to study the consequences of this inhibition on the role of calcium in circadian rhythmicity. This appears to be of special interest in the heart, known to display circadian rhythmicity (Reinberg, 1993).

3.7 Tumors

Transformed cells are characterized by the loss of cellular multiplication control and the progressive rise of new biochemicals. Altough literature abunds concerning glycoconjugates in cancer, the existence of biological rhythms is nearly ignored. Membrane glycoconjugates play a role in tumor growth and metastasis. Smaaland et al.(1993) observed that DNA synthesis and ploidy in lymphomas were dependent on circadian stage of cell sampling.

Some membrane components are probably released in the serum or in other biological fluids and serve as useful cancer markers as described in the next section.

3.7.1 Tumor markers

The term tumor marker addresses to both specific and non-specific markers, the specific markers being new structures appearing during the course of the disease. A number of tumor markers are membrane proteins or enzymes. One of the best studied membrane antigen is the carcinoembryonic antigen (CEA), a glycoprotein found mainly in the apical membranes of the colonic cells (Paynan and Reggio,1991). The CEA is anchored in the membrane by a glycosylphosphatidylinositol linkage (Garcia et al. 1991).

Another carbohydrate containing antigen, CA 15-3, is recognized by an antibody raised against milk fat globule membranes and by a second antibody raised against a membrane enriched fraction from breast carcinoma cells (Silver et al. 1991). CA 125 is present in ovarian tumors and in the serum of women with non-mucinous ovarian cancer (Kawabat et al. 1983).

The three markers display oscillations in the course of the 24-h scale. The amplitude may be as large as 200% demonstrating that a single determination per day is not sufficient (Touitou and Focan, 1994). The disappearance of the circadian rhythms was observed for CEA in the sera of cancer patients (Focan 1987). In some cases, the reappearance of the rhythm was observed after treatment (Touitou and Focan 1992).

In view of these results, a number of observations concerning molecules which seem to be modified with regard to their structures or their activities during the course of cancer diseases should be reconsidered, in particular, modified glycoconjugates and enzymatic activities which have been observed in the sera of some cancer patients. In order to refine the diagnosis and adapt treatment, several parameters -often used as cancer markers- such as blood group antigens (glycoconjugates), modified glycoproteins, glycolipids or glycosyltransferases should be studied in order to determine whether they undergo circadian variation and to test the hypothesis of a phase shift rather than a modified response level.

It should also be interesting to examine if glycosylation varies with the time of the day.

3.7.2 Chronotherapy

We have already mentioned the need for complementary studies on glycoconjugates rhythmicity in cancer. For a number of drugs, chronopharmacology represents an important step in the development of adequate treatments for cancer. An interesting question is the exact role played by the saccharidic units in the regulation of tumor growth and metastatic power . New drugs should control glycoconjugates metabolism or replace deficient glycoconjugates. Tunicamycin (inhibiting N-glycosylation) could alter rhythmicity (Galoutzov 1985 and Vanden Driessche 1992). Glycoproteins such as interleukins are potential powerful anticancer drugs and their delivery should be time-controlled. Balb/C mice developped cancer after inoculation with sarcoma cells. Injections of Interleukin-2 (IL-2), a cytokin carrying O-linked oligosaccharides, could accelerate or slow down tumor growth, depending on the time of treatment (during day), demonstrating a chronomodulation of the host-tumor balance (Sánchez de la Peña 1993). Marker rhythmometry should help clinicians in their cancer therapy (Elg et al. 1991).

3.8 Conclusion

Certain questions will arise with respect to the important diversity of glycoproteins due to the numerous isoforms themselves resulting from different glycosylations. Is this diversity time-dependent? Are the various isoforms potential rhythm modulators? These questions remain unsolved for membrane-bound as well as for free glycoproteins.

This diversity could be important for plasma membrane receptors. Therefore, it could be necessary to improve the analytical methods in order determine whether receptors carrying different oligosaccharides display a rhythmicity.

In some of the examples cited, the importance of terminal sugars was frequently reported. Sialic acids represent a polymorphic family of negatively charged carbohydrates. They are probably important for development (Varki 1993). Is the transfer of sialic acids by specific sialyltransferases relevant for the rhythm? The fixation of galactose, a neutral terminal sugar, may be essential for rod outer segment morphogenesis. Is the novel oligosaccharide detected in rhodopsin important for the circadian response?

Glycolipids or their derivatives, as modulators of cellular responses, are hypothetical second messengers exerting their effects on a number of important biochemical processes but their importance for circadian rhythms has not been determined.

Molecular biology offers interesting possibilities such as the modification of glycoproteins or that of the glycosyltransferases, the specific enzymes responsible for the synthesis of glycoconjugates, enabling us to study the molecules presumably involved in rhythmic operation.

Acknowledgments

The author thanks Galelephar SA. for its support.

References

Arendt J (1992) The pineal In Biological rhythms in clinical and laboratory medicine. Ed Touitou Y and Haus E, Springer-Verlag, Berlin, Heidelberg pp 348-362

Bargiello TA and Young MW (1984) Molecular genetics of biological clock in Drosophila. Proc Nat Acad Sci USA 81, 2142-2146

Baylies MK, Weiner L, Vosshall LB, Saez L, Young MW (1993) Genetic, molecular and cellular studies of the *per* locus and its products in Drosophila melanogaster. In: Young MW (ed) Molecular Genetics of Biological Rhythms. Marcel Dekker, New-York pp 123-153

Bernfield M (1989) Extracellular matrix. Curr Opin in Cell Biol 1,953-955

Bhavanadan VP and Davidson EA (1992) Proteoglycans: Structure, Synthesis, Function. In: Allen HJ and Kisailus EC (eds), Glycoconjugates. Marcel Dekker, New Yok pp167-202

Bick T, Youdim MBH and Hochberg Z(1989) Adaptation of liver somatogenic and lactogenic growth hormone (GH) binding to the spontaneous pulsations of GH secretion in the male rat.Endocrinology 125, 1711-1717

Bourdon MA, Oldberg A, Pierschbacher M and Ruoshlati E (1985) Molecular cloning and sequence analysis of a chondroitin sulfate proteoglycan cDNA. Proc Natl Acad Sci USA 82, 1321-132

Bourne HR, Sanders DA and McCormick F (1990) The GTPase superfamily : a conserved switch for diverse cell functions. Nature 348 ,125-132

Broquet P, Martin A, Peschard MJ, Baubichon-Cory H, Serres-Guillaume M and Louisot P (1987) Study of circadian correlations between acetylcholine muscarinic receptor and brain glycosyltransferase by multivariate analysis. Int J Biochem 19, 635-636

Caro JF, Cecchin F and Sinha MK (1984) Is glycosylation in the liver needed for insulin binding, processing, and action? Evidence of heterogeneity. J Biol Chem 259,12810-12816

Chory J (1994) Phytochrome signal transduction. Current Biol 4, 844-846

Cormier MJ, Anderson JM, Charbonneau H, Jones HP and McCann RO (1983) Calcium and Cell Function (Cheung WY ed) Academic Press, p201-218

Curatolo W(1987) Glycolipid function Biochim Biophys Acta 906,137- 160

Darvill AG, Albersheim P (1984) Phytoalexins and their elicitors -a defense against microbial infections in plants. Ann Rev Plant Physiol 35, 234-275

Darvill A, Augur C, Bergmann C, Carlson RW, Cheong JJ, Eberhard S, Hahn MG, Lo VM, Marfà V, Meyer B, Mohnen D, O'Neill MA, Spiro MD, van Halbeek H, York WS and Albersheim P (1992) Olicosaccharins -Oligosaccharides that regulate growth, development and defence responses in plants. Glycobiol 2, 181-198

Driouich A, Gomord V, Fichette-Lainé AC and Faye L(1993) Glycosylation des protéines végétales sécrétées et biosynthèse des glycannes complexes. Regards sur la Biochimie 3, 33-42

Drobny EC, Amburn K and Baumann G(1983) Circadian variation of basal plasma growth hormone in man J Clin Endocrinol Metab 57, 524-528

Dryer SE and Henderson D (1991) A cyclic-GMP activated channel indissociated cells of the chick pineal gland. Nature 353 ,756-758

Duffin KL, Lange GW, Welply JK, Florman R, O'Brien PJ, Dell A, Reason AJ, Morris HR and Fliesler SJ (1993) Identification and oligosaccharide structure analysis of rhodopsin glycoforms containing galactose and sialic acid. Glycobiol 3, 365-380

Dull TJ , Gray A, Hayflick JS and Ullrich A(1984) The insulin-like growth factor II precursor gene organization in relation to insulin gene family. Nature, 310, 777-781

Dunlap JC(1993) Genetic analysis of circadian clocks. Ann Rev Physiol 55, 683-728

Duronio V, Jacobs S and Cuatrecasas P(1986) Complete glycosylation of the insulin and the insulin-like growth factor I receptors is not necessary for their biosynthesis and function. Use of swainsonine as an inhibitor in IM-9 cell. J Biol Chem 261,970-975

Duronio V, Jacobs S, Romero PA and Herscovics A (1988) Effects of inhibitors of N-linked oligosaccharide processing on the biosynthesis and function of insulin and insulin-like growth factor I receptors. J Biol Chem 263, 5436-5445

Edmunds LN Jr (1988) In:Cellular and molecular Bases of Biological clocks.Springer-Verlag, Heidelberg, Berlin

Edmunds LN Jr (1992) Cellular and molecular aspects of circadian oscillators: Models and mechanisms for biological time keeping. In Biological rhythms in clinical and laboratory medicine. Ed Touitou Y and Haus E, Springer-Verlag, Berlin, Heidelberg pp 35-54

Edmunds LN Jr and Tamponnet C (1990) Oscillator control of cell division cycles in Euglena: role of calcium in circadian time keeping. In Calcium as an intracellular messenger in Eucaryotic microbes. O'Day DH(ed) Am Soc for Microbiol, Washington DC USA, 97-123

Elg S, Halberg E, Ramakrishnan S, Cornélissen G, Haus E, Nicolau G, Carson L, Twiggs L,Long III HJ and Halberg F(1991)Marker rhythmometry with macrophage-colony stimulating factor (M-CSF) Chronobiologia 18, 141-152

Focan C (1987) Chronobiologie et marqueurs biochimiques du cancer humain. Path Biol 35, 951-959

Foulkes NS, Schlotter F, Pévet P and Sassone-Corsi P (1993) Pituitary hormone FSH directs the CREM functional switch during spermatogenesis. Nature 362, 264-267

Fujita S, Endo T, Ju J, Kean EL and Kobata A (1994) Structural studies of the N-linked sugar chains of human rhodopsin. Glycobiol 4,633-640

Fukui H (1989) Biological activities of saccharides in plant cells. Trends in Glycosc and Glycobiol, 1, E02- E08

Galoutzov B (1985) Biological and biophysical effects of monensin and tunicamycin in Acetabularia acetabulum In:Acetabularia 1984 Ed Bonotto S, Cinelli F and Billiau R Belgian Nuclear Center, Mol 169-182

Garcia M, Seigner C, Bastid C, Choux R, Payan MJ and Reggio H (1991) Carcinoembryonic antigen has a different molecular weight in normal colon and in cancer cells due to N-glycosylation. Cancer Res 51, 5679-5686

Hakomori SI (1981) Glycosphingolipids in cellular interaction, differentiation and oncogenesis. Ann Rev Biochem 50, 733-764

Hakomori SI (1990) Bifunctionnal role of glycosphingolipids . J Biol Chem 265, 18713-18716

Hannun YA and Bell RM (1989) Function of sphingolipids and sphingolipid breakdown products in cellular regulation. Science 243, 500-507

Hargrave PA and McDowell JH (1992) Rhodopsin and phototransduction. Int Rev Cytol 137B, 49-97

Hilbush BS and Levine JM (1992) Modulation of a Ca^{2+} signaling pathway by GM1 ganglioside in PC 12 Cells . J Biol Chem 267, 24789-24795

Igarashi Y, Nojiri H, Hanai N and Hakomori SI (1989) The gangliosides that modulate membrane protein function. Methods in Enzymol 179, 521-541

Ju JM and Kean EL (1992)In vitro galactosylation of rhodopsin and opsin:kinetics, properties and characterization. Exp Eye Res 55 ,589-604

Julius D, Mac Dermott AB, Jessel TM, Huang K, Molineaux S, Schieren I and Axel R (1988) Fuctionnal expression of the 5-HT_{1c} receptor in neuronal and nonneuronal cells. Cold Spring Harbor Symp 53, 385-393

Kandel ER, Abrams TW, Bernier L, Carew TJ, Hawkins RD and Schwartz JH(1983) Classical conditioning and sensitization share aspects of the same molecule cascade in Aplysia. Cold Spring Harbor Symp Quant Biol 48,821-830

Karlsson KA (1977) Aspects on structure and function of sphingolipids in cell surface membranes. In Structure of Biological Membranes Ed Abrahamson S and Pascher I, Plenum Press New York pp 245-274

Kaupp UB, Hanke W, Simmoteit R and Lehring H (1988) Electrical and biochemical properties of cGMP-gated cation channel from rod photoreceptors. Cold Spring Harbor Symp 53,407-417

Kawabat SE, Bast RC, Welsh WR, Knapp RC and Calvin RB (1983)Immunopathologic characterization of monoclonal antibody that recognizes common surface antigens of a human ovarian tumor of serous, endometroid and clear cell types. Am J Clin Pathol 79,98-104

Klausen T, Flemming D, Hippe E and Galbo H (1993) Diurnal variation of serumerythropoietin in trained and untrained subjects.Eur J Appl Physiol 67,545-548

Klevecz RR and Bolen JJ Jr (1993) Dynamic analysis of Period mutants in Drosophila: How a precize circadian clock might emerge from a tissue composed of chaotic cellular oscillators. In Molecular genetics of biological rhythms.Ed Young MW, Marcel Dekker New-York pp221-253

Korenbrot JI and Fernald RD (1989) Circadian rhythm and light regulated opsin mRNA in rod photoreceptors. Nature 337, 454-457

Koumenis C and Eskin A (1992) The hunt for mechanisms of circadian timing in the eye of Aplysia Chronobiol Intern 9, 201-221

Kundu SK (1992) Glycolipids:Structure,Synthesis and Function. In Glycoconjugates,ed Allen HJ and KisailusEC, Marcel Dekker Inc New York p 203-262

Lascols O, Caron M, Capeau J, Cherqui G and Picard J (1983) Structural carbohydrates in the liver, Falk Symposium 34, Ed Popper H Reutter,Gudat F and Kottgen,p 635, MTP Press, Boston,The Hague, p 635

Lewandowski MH (1988) Seasonal variations in the circadian activity of Acetylcholinesterase in the brain stems reticular formation of mice under normal and constant light regimen. Chronobiol Intern 5, 121-125

Lloyd D and Lloyd AL (1993) Hypothesis: A controlled chaotic attractor constitutes the central oscillator of the circadian clock. Biochem Soc Transaction 22, S322

Louisot P (1982) The biosynthesis of glycoconjugates and its control. Progr Clin Biol Res 102, ptA 385-395

Lumsden PJ (1991) Circadian rhythms and the phytochrome Ann Rev Plant Physiol Plant Mol Biol 42, 351-371

Marmé D and Dieter P (1983) : Calcium and Cell Function, Ed Cheung WY Academic Press, vol IV ,264-311

McClung CR, Fox BA and Dunlap JC (1989) The Neurospora clock gene frequency shares a sequence element with the Drosophila clock gene period. Nature,339, 558-562

McKeon JL, Saunders NA, Murree-Allen K, Olson LG, Gyulay S, Dickeson J, Houghton A, Wlodarczyk J and Hensley MJ (1990) Urinary uric acid:creatinine ratio,serum erythropoietin, and blood 2,3-diphosphoglycerate in patients with obstructive sleep apnea. Am Rev Respir Dis 142,8-13

Meier AH (1973) Daily hormone rhythms in the white-throated sparrow. Am Sci 61, 184-187

Miller ME, Garcia JF, Cohen RA, Cronkite EP, Moccia G and Acevedo J (1981) Diurnal levels of immunoreactive erythropoietin in normal subjects and subjects with chronic lung disease. Brit J of Haematol 49,189-200

Mode A, Gustafsson JA, Jansson JO, Eden S and Isaksson OGP (1982) Association between level of growth hormone and sex differentiation of hepatic steroid metabolism. Endocrinology 11, 1692-1697

Neuhaus G, Bowler C, Kerr R, Chua NH (1993) Calcium/Calmodulin-dependent and independent phytochrome signal transduction pathways. Cell 93, 937-952

Nomura K, Tsunasawa S, Ohmura K, Sakiyama F and Shizume K (1988) Renotopic activity in ovine luteinizing hormone isoform(s). Endocrinology 123, 700-712
Okano T, Yoshizawa T and Fukada Y (1994) Pinopsin is a chicken pineal photoreceptive molecule. Nature,372,94-97
Okazaki T, Domae N, Bell RM, Hannun Y (1994) Ceramide, a glycolipid second messenger. Trends in Glycosc and Glycotechnol 6, 278-285
Padmanabhan V, Lang LL, Sonstein J, Kelch RP and Beitins IZ (1988) Modulation of serum follicle-stimulating hormone bioactivity and isoform distribution by estrogenic steroids in normal women and in gonadal dysgenesis. J of Clin Endocrin and Metab 67, 465-473
Paynan MJ and Reggio H (1991) Carcinoembryonic antigen has a different molecular weight in normal colon and in cancer cells due to N-glycosylation differences. Cancer Res 51, 5679-5686
Purushotham KR, Bologna J, Nakagawa Y and Humphreys-Beher MG (1992) Isolation and characterization of a new Ca^{2+}/calmodulin dependent protein kinase from isoproterenol-stimulated proliferating rat parotid acinar cells. Biochem Cell Biol 70, 250-255
Reinberg A (1993) Les rythmes biologiques. Coll Que sais-je? Presses Universitaires de France, Paris, pp128
Riboni L and Tettamanti G (1993) Ganglioside turnover and formation of metabolic regulators. Trends Glycosci Glycobiol 5, 1-1
Riov J and Jaffe M (1973) Cholinesterase from plant tissues. Plant Physiol 51, 520-528
Roberts K (1989) The plant extracellular matrix.CurOpin Cell Biol. 1,1020-1027
Ruoslahti E (1988) Structure and biology of proteoglycans. Ann Rev Cell Biol 4, 229-255
Runyan RB, Versalovic J and Shur BD (1988) Functionnaly distinct laminin receptors mediate cell adhesion and spreading: the requirement of surface galactosyltransferase in cell spreading. J Cell Biol 257,9740-9744
Ryan CA (1987) Oligosaccharide signalling in plants. Ann Rev Cell Biol 3, 295-317
Sánchez de la Peña S, Hrushesky WJM, Wood P, Vysula R (1993) Host-tumor balance depends upon IL-2 Circadian timing.Proc Am Ass Cancer Res 340 ,57
Satiat-Jeunemaitre B (1992) Spatial and temporal regulations in helicoidal extracellular matrices: comparison between plant and animal systems.Tissue and Cell 24,315-334
Schachner M (1989) Families of neural adhesion molecules,Carbohydrate recognition in cellular function,Ciba foundation Symposium 145,Ed John Wiley & sons, pp 156-172
Schweiger HG and Schweiger M (1977) Circadian rhythms in unicellular organisms: an endeavor to explain the molecular mechanism. In Intern Rev of Cytol , Ed Bournes GH, Danielli JF and Jeon KW vol51 pp315-342
Sharrock RA and Quail PH (1989) Novel phytochrome sequences in Arabidopsis thaliana: structure, evolution, and differential expression of a plant regulatory photoreceptor family. Genes & Development 3, 1745-1757
Silver HKB, Archibald BL, Ragaz J, Coldman AJ (1991) Relative operating characteristic analysis and group modeling for tumor markers: comparison of CA 15-3, carcinoembryonic and mucin-like carcinoma associated antigen in breast cancer carcinoma. Cancer Res 51, 1904-1909
Smaaland R, Lote K, Sothern RB and Laerum OD (1993) DNA synthesis and ploidy in non-Hodkin's Lymphomas demonstrate intrapatient variation depending on Circadian Stage of cell sampling. Cancer Res 53, 3129-3138
Smith SB, St Jules RS and O'Brien PJ (1991) Transient hyperglycosylation of rhodopsin with galactose. Exp Eye Res 53,525-537
Stehle JH, Foulkes NS, Molina CA, Simonneaux V, Pévet P and
Sassone-Corsi P (1993) Adrenergic signals direct rhythmic expression of transcriptional repressor CREM in the pineal gland. Nature 365 ,314-320
Takahashi JS (1994) ICER is nicer at night(sir!) Current Biology 4 ,165-168

Takahashi JS and Zatz M (1982) Regulation of circadian rhythmicity. Science 217,1104-1111

Takeuchi M and Kobata A (1991) Structures and functional roles of the sugar chains of human erythropoietins. Glycobiol 1,337-34

Touitou Y and Focan C (1992) Rhythms in tumor markers . In: Biological Rhythms in clinical and laboratory medicine. Ed Touitou Y and Haus E, Springer-Verlag pp 648-657

Tretyn A and Tretyn M (1990) Oscillations of acetylcholine in oat seedlings. Chronobiologia 17,45-52

Uehara F, Matthes MT,Yasumura D and LaVail MM (1990) Light-evoked changes in the interphotoreceptor matrix. Science 248, 1633-1636

Ujihara M, Yamamoto K, Nomura K, Toyoshima S, Demura H, Nakamura Y,Ohmura K and Osawa T (1992) Subunit-specific sulphation of oligosaccharides relating to charge heterogeneity in porcine lutrophin isoforms. Glycobiol 2, 225-231

Vanden Driessche T, Dujardin E, Magnusson A and Sironval C (1976) Acetabularia mediterranea: circadian rhythm of photosynthesis and associated changes in molecular structure of the thylakoid membrane. Int J Chronobiol 4, 111-124

Vanden Driessche T (1986) Polarity and Circadian Rhythmicity are regulatory components in Acetabularia morphogenesis. Endocyt Res 3, 275-297

Vanden Driessche T, Lause P, Guisset JL, Glory M and Petiau-de Vries GM(1992) Some aspects of glycoprotein metabolism in Acetabularia: Spatio-temporal activity of β-galactosidase and physiological effects of tunicamycin. J Interdiscipl Cycle Res 23, 237-251

Vanden Driessche T (1994) Circadian Rhythms in three unicellular organisms. The peculiarities of the organisms, the evidence they brought on rhythms and their specific practical problems. Outline of recent hypotheses. Biol Rhythm Res 25, 345-386

Varki A (1993) Biological role of oligosaccharides:all the theories are correct.Glycobiol 3, 97-130

Veldhuis JD, King JC, Urban RJ, Rogol AD, Evans WS, Kolp LA and Johnson ML (1987) Operating characteristics of the male hypothalamo-pituitary-gonadal axis: pulsatile release of testosterone and follicle-stimulating hormone and their temporal coupling with luteinizing hormone. J of Clin Endocrin and Metab 65,929-941

Warner MD, Sinha YN and Peabody CA (1993) Growth hormone and prolactin variants in normal subjects.Relative proportions in morning and afternoon samples. Horm Metab Res 25, 425-429

Wide L, Bengtsson C and Birgegard G (1989) Circadian rhythm of erythropoietin in human serum. Brit J of Haemathol 72, 85-90

Witz P,Amlaiki N, Plassat JL, Maroteaux L, Borrelli E and Hen R (1990) Cloning and characterization of a Drosophila serotonin receptor that activates adenylate cyclase. Proc Natl Acad Sci USA 87, 8940-8944

Yamaguchi Y (1993) Proteoglycan-growth factor interactions. Trends in Glycosci Glycotechnol 5, 428-437

Younes A, Kahn DW, Besterman JM, Bittman R, Byun H, and Kolesnick RN (1992) The evidence of the functioning of ceramide as second messenger. J Biol Chem 267, 842-847

Yee HFJr, Kuwata JH and Langer GA (1991) Effects of neuraminidase on cellular calcium and contraction in cultured cardiac myocytes. J Mol Cell Cardiol 23, 175-185

Zehring WA, Wheeler DA, Reddy P,Konopka RJ, Kyriacou CP, Rosbash M and Hall JC (1984) P-element transformation with periodic locus DNA restores rhythmicity to mutant arrhythmic Drosophila melanogaster. Cell 39, 369-376

4 From Ultradian Biochemical Oscillations to Circadian Rhythms

Albert Goldbeter
Unité de Chronobiologie Théorique, Service de Chimie Physique, Faculté des Sciences, Université Libre de Bruxelles, Campus Plaine, C.P. 231, B-1050 Brussels, Belgium

Summary

The goal of this chapter is to examine some of the best-known examples of ultradian biochemical oscillations and to draw a comparison between these periodic phenomena and circadian rhythms. Successively considered in the first part of the chapter are (i) glycolytic oscillations observed in yeast and muscle, (ii) the periodic synthesis of cyclic AMP (cAMP) signals which govern aggregation and differentiation in the cellular slime mold Dictyostelium discoideum, and (iii) the mitotic oscillator that controls the early cell division cycles in amphibian embryos. For each of these biochemical rhythms, the salient experimental aspects and the insights provided by theoretical models are summarized in turn. The second part of the chapter is devoted to the mechanism of circadian rhythms in Drosophila. A theoretical model for circadian oscillations in the *period* protein (PER) is presented. This minimal model is based on multiple phosphorylation of PER and on the negative feedback exerted by PER on the transcription of the *period* gene. Parallels are drawn between the mechanisms underlying ultradian biochemical oscillations and the mechanism that governs the circadian, periodic variation of PER in Drosophila.

4.1 Introduction

In searching for the molecular mechanism of circadian rhythmicity, it is useful to consider these oscillatory phenomena in the general framework of biological rhythms which occur, with widely different periods, at all levels of biological organization. Of particular interest among these are biochemical oscillations produced by oscillatory enzymes or oscillatory networks of enzyme reactions. The present chapter aims at drawing a comparison between these ultradian biochemical oscillations and circadian rhythms. Examined in turn for each of these rhythms will be the salient experimental features and the insights provided by theoretical models. Models help to clarify the mechanism of oscillations as they contribute to make clear how regulatory interactions at the molecular level give rise to sustained oscillations in biological systems.

Rather than providing a survey of biochemical oscillations and other ultradian rhythms, which have been reviewed elsewhere (Hess and Boiteux 1971; Goldbeter and Caplan 1976; Berridge and Rapp 1979; Lloyd and Stupfel 1991; Goldbeter 1995a), we focus in the first part of this chapter on three of the best-known examples of ultradian biochemical oscillations whose mechanism is known in some detail. Considered in Section 2 are (i) glycolytic oscillations observed in yeast and muscle, (ii) the periodic synthesis of cyclic AMP (cAMP) signals which govern aggregation and differentiation in the cellular slime mold Dictyostelium discoideum, and (iii) the mitotic oscillator that controls the eukaryotic cell division cycle.

Glycolytic and cAMP oscillations have a period of several minutes, whereas the period of the mitotic oscillator increases from about 30 min for the division of amphibian embryonic cells up to 24 hours or more for the division of somatic cells, depending on the cell type. This brings us to the period range of circadian rhythms whose molecular bases are discussed in Section 4.3 (the cell cycle clock in fact identifies with a circadian oscillator in several organisms such as Euglena; see the chapter by Edmunds in this volume). In addressing the mechanism of circadian rhythmicity, we focus on the case of circadian oscillations of the *period* protein (PER) and its mRNA in Drosophila, since this system is presently the best characterized at the molecular level. Based on experimental observations, we present a minimal model for circadian oscillations of PER in Drosophila. The link between this model and models previously proposed for ultradian biochemical oscillations is emphasized. Not directly dealt with, but alluded to, is the intriguing observation that upon disruption, circadian rhythms sometimes disappear to reveal the existence of ultradian oscillations (see, for example, Dowse and Ringo 1993).

Section 4.4 will be devoted to concluding remarks and to a comparison of the various rhythms considered. A common point, suggested by the study of theoretical models, is that they all correspond to sustained oscillations of the limit cycle type. More specific analogies can be drawn, according to several criteria. Thus, whereas glycolytic and cAMP oscillations rely on positive feedback, the PER circadian oscillator and the mitotic oscillator appear to rely, respectively, on negative feedback and on a combination of negative and positive regulations. Two of the rhythms involve membrane processes: a plasma membrane receptor plays a central role in cAMP oscillations in Dictyostelium, whereas PER circadian oscillations involve the transport of the PER protein and its mRNA across the nuclear membrane. Protein synthesis and degradation are involved in mitotic and circadian oscillations. Another parallel pertains to the role of protein phosphorylation: the cAMP oscillator in Dictyostelium and the mitotic oscillator in eukaryotic cells both rely on regulation by reversible phosphorylation, which also contributes to circadian oscillations of PER in Drosophila.

4.2 Biochemical Oscillations

4.2.1 Glycolytic Oscillations

Glycolytic oscillations are the prototype for periodic phenomena in biochemistry, even if the function of these oscillations, in many cases, remains unclear. The phenomenon consists in the periodic transformation of a sugar, injected at a constant rate, by the enzymes that constitute the glycolytic pathway. Glycolytic oscillations have mostly been studied in yeast cell suspensions or extracts, and muscle extracts. Renewed interest in the phenomenon stems from the recent observation of glycolytic oscillations in both pancreatic β-cells (Chou et al. 1992) and heart cells (O'Rourke et al. 1994).

Experimental Observations on Glycolytic Oscillations. The initial observations by Chance and coworkers (Chance et al. 1964; Ghosh and Chance 1964; Pye and Chance 1966) showed that glycolytic oscillations could be maintained in yeast suspensions for relatively long periods of time. Shortly thereafter observations were conducted on yeast extracts, with very similar results. Regular oscillations were recorded for many periods, provided that the glycolytic substrate was produced in a continuous manner (Chance et al. 1964; Pye 1969). The periodic phenomenon is also observed in a single yeast cell (Chance et al. 1967). Recording the fluorescence of NADH allows the continuous measurement of the phenomenon, which can also be studied by chemically assaying the concentrations of glycolytic intermediates or following changes in pH (Hess et al. 1969).

The essential property of glycolytic oscillations, demonstrated in yeast extracts (Hess and Boiteux 1968; Hess et al. 1969) and yeast cell suspensions (Von Klitzing and Betz 1970), is that sustained periodic behavior is observed only in a range of substrate injection rate. Below a critical value of the substrate injection rate, the system reaches a stable steady state. When this rate increases, oscillations occur, but they disappear when the substrate injection rate exceeds a second, higher critical value. The period of glycolytic oscillations diminishes from about 8 to 3 min as the substrate injection rate increases from some 20 to 160 mM/h (Hess et al. 1969).

Very early on, the source of oscillations within the glycolytic system was identified (Ghosh and Chance 1964; Hess and Boiteux 1968, 1971; Pye 1969). Glycolysis represents a chain of enzyme reactions which in yeast transforms a sugar such as glucose or fructose into ethanol and CO_2. When a hexose such as glucose 6-phosphate or fructose 6-phosphate is taken as glycolytic substrate, periodic behavior is observed. This indicates that the source of oscillations lies beyond the first two enzymes of the chain, hexokinase and glucose-phosphate isomerase. However, when the phosphofructokinase (PFK) step is bypassed by injecting fructose 1,6-bisphosphate as glycolytic substrate, oscillations disappear. Periodic behavior therefore originates in the reaction catalyzed by PFK.

The role of PFK is corroborated by the fact that inhibitors or activators of the enzyme affect periodic behavior. This is the case, for example, for ammonium ions that activate yeast PFK; their addition suppresses the oscillations (Hess and Boiteux 1968). In muscle, where glycolytic oscillations are also observed (Frenkel 1968; Tornheim et al. 1991), the molecular mechanism is identical: the enzyme responsible for the phenomenon is again PFK. Citrate, an inhibitor of PFK, controls the oscillations in muscle extracts (Frenkel 1968; Tornheim et al. 1991). Moreover, as in yeast, the addition of purified PFK produces a phase shift of the oscillations and permits modulation of their period and amplitude; above a threshold, the addition of PFK suppresses the oscillations, but these reappear when the system is supplied with hexokinase (Frenkel 1968). The latter observation, together with the effect of the substrate injection rate, indicates that periodic behavior in glycolysis depends on a delicate balance between processes producing and transforming the substrate in the reaction catalyzed by PFK.

What are the characteristics that confer on PFK the capability of producing oscillations? The distinctive property of the enzyme was rapidly identified (Ghosh and Chance 1964): PFK is activated by a reaction product, ADP, and by AMP, which is converted into ADP in the reaction catalyzed by adenylate kinase. The control exerted by ADP is readily demonstrated by the immediate phase shift induced upon addition of this metabolite (Hess and Boiteux 1968; Pye 1969). The activation of an enzyme by its product is relatively rare in biochemistry. Regulation by negative feedback is much more common, as exemplified by the inhibition of an enzyme by the end product of a metabolic chain (how such negative feedback can produce oscillations will be discussed below). Glycolysis does not provide the only example of an enzyme activated by its reaction product: another case of autocatalytic regulation of enzyme activity, considered in Section 4.2.2, is that of adenylate cyclase in Dictyostelium amoebae.

Insights from Theoretical Models for Glycolytic Oscillations. How the auto--catalytic regulation of PFK leads to glycolytic oscillations is a question that benefits from being put in theoretical terms. The knowledge of the mechanism of oscillations and the availability of numerous experimental data early on prompted the construction of models for the autocatalytic PFK reaction (Higgins 1964; Sel'kov 1968). A model taking into account explicitly the allosteric nature of PFK was later developed (Goldbeter and Lefever 1972; Goldbeter and Nicolis 1976; see also Goldbeter, 1995a). This model considers an allosteric enzyme activated by the reaction product; the two variables of this autocatalytic biochemical system are the substrate and product concentrations (Fig.4.1, part A). As in the experiments described above, the substrate is injected at a constant rate, v. The allosteric enzyme contains several subunits, which exist in two conformations, one active (R) and the second (T) less active or inactive; the transition of all subunits between the two states is concerted. In the absence of substrate and product, the enzyme is predominantly in the T state. Activation by the product occurs when the latter binds to a regulatory site of the enzyme in the R

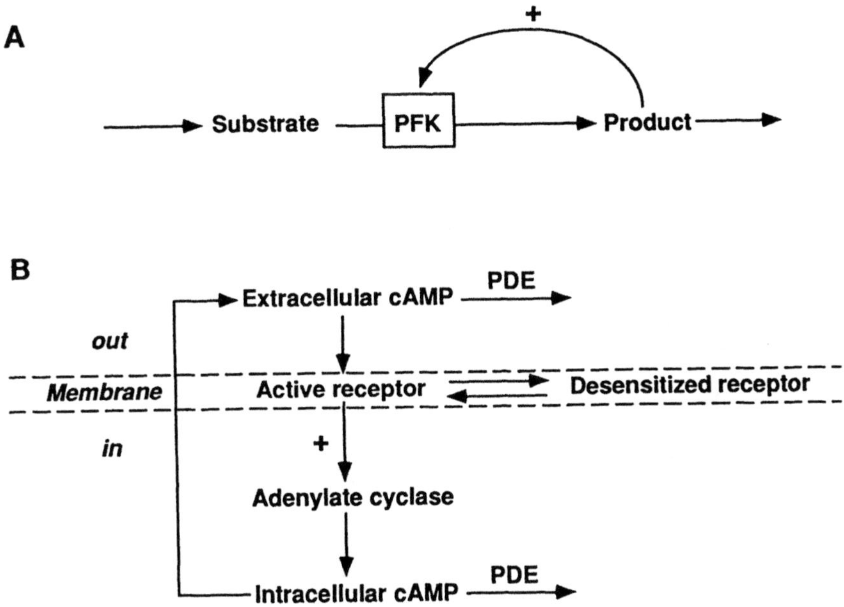

Fig.4.1 Schematic representation of the mechanisms responsible for some of the best-known examples of ultradian biochemical oscillations based on positive feedback. (A) The product-activated phosphofructokinase (PFK) reaction responsible for glycolytic oscillations in yeast and muscle. (B) In the slime mold Dictyostelium discoideum, the mechanism of cAMP oscillations involves the activation of adenylate cyclase upon binding of cAMP to the active form of the cAMP receptor, desensitization of the cAMP receptor through reversible phosphorylation, transport of intracellular cAMP into the extracellular medium, and cAMP hydrolysis by phosphodiesterase (PDE). The positive feedback loop apparent in both cases underlies the mechanism of sustained oscillations.

conformation, and thereby shifts the equilibrium of the enzyme forms toward the active state.

The analysis of the model shows that sustained oscillations occur in a range bounded by two critical values of the substrate injection rate. In that range, in agreement with experimental observations, the period of the oscillations decreases as the substrate input rate rises. In the phase plane, when the substrate concentration is plotted as a function of the product concentration in the course of time, the system evolves towards a unique closed curve, regardless of initial conditions (Fig.4.2). Such a curve, known as a limit cycle, corresponds to highly stable oscillations characterized by a unique amplitude and frequency for a given set of parameter values, irrespective of transient perturbations applied to the oscillatory system.

This simple, two-variable model of a product-activated enzyme reaction suffices to account for many observations on glycolytic oscillations in yeast and muscle (Boiteux et al. 1975; Goldbeter and Caplan 1976; Goldbeter 1995a). More

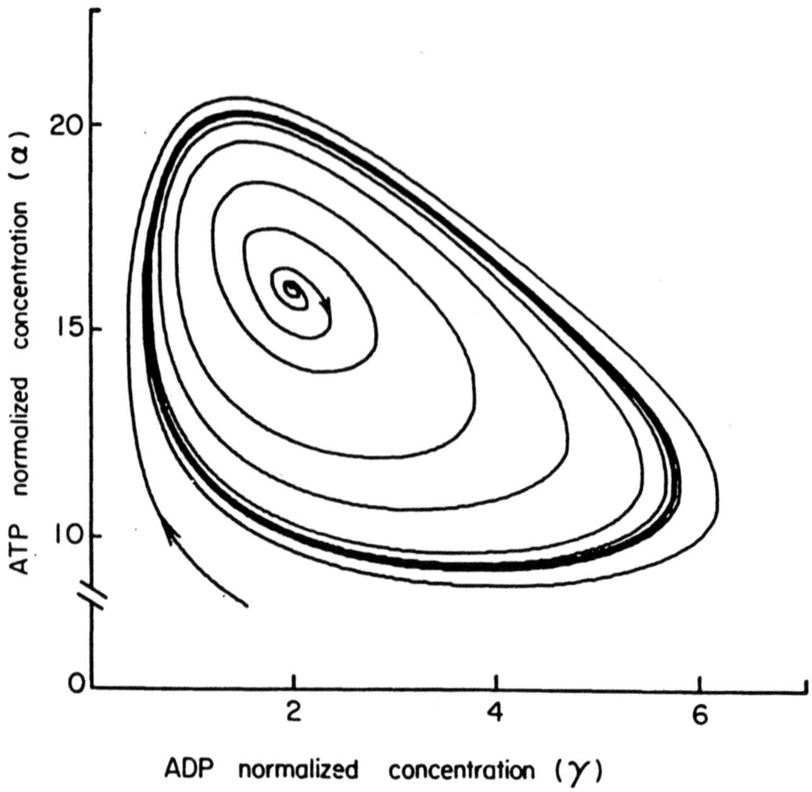

Fig.4.2 Limit cycle generated by the allosteric model for the product-activated phosphofructokinase reaction responsible for glycolytic oscillations (Goldbeter and Caplan 1976)

complex models have been proposed, which take into account additional reactions of the glycolytic pathway but remain centered on the PFK autocatalytic step. As in the simple two-variable model, these more complex models predict the occurrence of sustained oscillations of the limit cycle type. Glycolytic oscillations thus provide us with an example of ultradian biochemical rhythm due to a key enzyme reaction regulated by positive feedback.

4.2.2 Cyclic AMP Oscillations in Dictyostelium

The amoeba Dictyostelium discoideum is one of the most studied organisms in developmental biology (Bonner 1967; Devreotes 1989). These cells can switch from a unicellular to a multicellular stage during their life cycle by resorting to a chemical mechanism of intercellular communication (Gerisch 1987; Devreotes 1989), which presents some similarities with hormonal communications in higher organisms. Once the multicellular stage is reached, the amoebae differentiate into at least two distinct cell types, thus providing a simple model for the study of pattern formation and cell differentiation in eukaryotes. It is for yet another reason that Dictyostelium amoebae draw our attention here. The mechanism of intercellular communication that governs the transition between the isolated and collective phases of their life cycle possesses a periodic nature. After reviewing the experimental facts that make this phenomenon a prototype for biochemical oscillations in intercellular communication, theoretical models for this rhythm will briefly be discussed.

Experimental Aspects of cAMP Oscillations. Growth and division of D. discoideum amoebae proceed until food becomes scarce. Following starvation, the amoebae collect to form an aggregate that may contain up to 10^5 cells; the resulting aggregate takes the form of a slug, and finally differentiates, after a phase of migration, into a fruiting body, which consists of a mass of spores surmounting a stalk that can measure up to a few mm. Spore dispersal leads back to the initial phase of the life cycle (Bonner 1967; Gerisch 1987; Devreotes 1989).

The aggregation of D. discoideum cells on agar proceeds in a periodic manner. Time lapse films of aggregation demonstrate the occurrence of waves of amoebae moving with a periodicity of the order of 5 to 10 min (Durston 1974). The amoebae collect around cells behaving as aggregation centers that periodically emit a chemical signal to attract the cells present in their environment. Amoebae are capable of relaying the signals emitted by a center located in their vicinity. This explains the wavelike nature of aggregation over territories whose dimensions can reach up to 1 cm: within each aggregation territory, the amoebae move in concentric or spiral waves (Alcantara and Monk 1974; Durston 1974). The chemotactic signal for D. discoideum was shown to be cAMP. Waves of cellular movement correlate with waves of cAMP; the latter present a striking analogy with waves observed in oscillatory chemical systems such as the Belousov-Zhabotinsky reaction (Winfree 1972).

An important step in the characterization of cAMP oscillations was made when Gerisch and Hess (1974) studied the behavior of D. discoideum amoebae in continuously stirred suspensions. By recording light scattering within such cell suspensions, these authors showed that spontaneous oscillations occur several hours after the beginning of starvation, with a period of the order of 5 to 10 min. The periodicity of oscillations in suspensions thus compares with that of the waves observed during aggregation on agar. The initial study of Gerisch and Hess (1974) further showed that extracellular cAMP is capable of inducing an immediate phase shift of the oscillations.

The existence of cAMP oscillations concomitant with the periodic variation in light scattering was confirmed shortly thereafter by Gerisch and Wick (1975). Intracellular cAMP oscillates in phase with light scattering, with a period close to 9 min; its peak precedes by some 10 s that of extracellular cAMP. The oscillations in intra- and extracellular cAMP can also be phase shifted by a pulse of extracellular cAMP (Gerisch et al. 1979). The effect of the perturbation depends on the phase at which the addition of cAMP occurs: a delay of the next peak is obtained when the stimulation takes place just after a peak, when cAMP is still close to maximum; thereafter the stimulation gives rise to the synthesis of a precocious peak of cAMP. Combining these results on phase shift as a function of phase of perturbation yields a phase response curve characteristic of a number of biological oscillators, including circadian rhythms.

Insights from Theoretical Models for cAMP Oscillations. As for glycolytic oscillations, the mechanism of cAMP oscillations in Dictyostelium relies on positive feedback (Fig.4.1, part B): upon binding to a plasma membrane receptor, extracellular cAMP elicits the synthesis of intracellular cAMP. To avoid a runaway phenomenon, however, such amplification has to induce its own limitation. For glycolytic oscillations, the limitation of autocatalysis originates from substrate consumption, which follows the activation of PFK by its reaction product. In D. discoideum, the limitation is due to cAMP-induced desensitization of the cAMP receptor through reversible phosphorylation (Devreotes and Sherring 1985). The receptor is dephosphorylated in a few minutes after removal of the external stimulus. Desensitization arises because the phosphorylated form of the cAMP receptor is less apt at promoting the synthesis of intracellular cAMP. In the course of cAMP oscillations, the receptor varies periodically between the phosphorylated and dephosphorylated states (Gundersen et al. 1989).

A model for cAMP oscillations in Dictyostelium based on the reversible desensitization of the cAMP receptor has been proposed (Martiel and Goldbeter 1987; Goldbeter 1995a). In this model, extracellular cAMP binds to the receptor, which exists in two states, one of which is active (R) and the other desensitized (D); only the complex formed by cAMP with the receptor in the R state is capable of activating adenylate cyclase, the enzyme synthesizing cAMP. The positive feedback loop arises from the transport of intracellular cAMP into the extracellular medium where it binds to the cAMP receptor and is hydrolyzed by phosphodiesterase. This model accounts for the oscillatory synthesis of the cAMP signal, with a periodicity of 5 to 10 min, and for the accompanying, periodic alternation of the receptor between the phosphorylated (D) and dephosphorylated (R) states. The model predicts that the interval between two cAMP peaks - and, consequently, the period of the oscillations - is primarily set by the time required for resensitization by the phosphatase that dephosphorylates the cAMP receptor.

Besides accounting for cAMP oscillations, the model for cAMP signaling based on receptor desensitization accounts for the relay of suprathreshold pulses of cAMP, which allows centers to control the chemotactic movement of a large number of amoebae over territories of macroscopic dimensions during aggregation. The model also provides an explanation for the sequence of developmental transitions *no relay - relay - oscillations* observed during the hours that follow starvation (Gerisch et al. 1979). The theoretical analysis shows that the continuous increase in the activities of adenylate cyclase and phosphodiesterase after starvation suffices to account for these developmental transitions (Goldbeter and Segel 1980; Goldbeter and Martiel 1988; Goldbeter 1995a). In this view, aggregation centers would be those cells that would be the first to reach the domain of sustained, autonomous oscillations of cAMP as a result of a progressive rise in the activities of adenylate cyclase and phosphodiesterase. Such a phenomenon provides a prototype for the ontogenesis of biological rhythms, by showing how continuous changes in biochemical parameters - associated here with the synthesis of enzymes or receptors - can lead to the onset of autonomous oscillations of the limit cycle type once a critical value of a control parameter is exceeded.

4.2.3 The Mitotic Oscillator

With regard to the physiological role of biochemical oscillations, one of the most important examples is provided by the mitotic oscillator that controls the onset of cell division in eukaryotes (Murray and Hunt 1993). This biochemical oscillator, driven by the synthesis of proteins named cyclins, governs the periodic activation of an enzyme known as cdc2 kinase. Activation of the latter enzyme in a cascade of phosphorylation-dephosphorylation cycles triggers the onset of mitosis. Such a mechanism appears to be universal among eukaryotes (Nurse 1990), although it takes its simplest form in the early stages of amphibian embryogenesis (Murray and Kirschner 1989a, b; Murray and Hunt 1993). Whereas in somatic cells and yeast the cell cycle consists in a succession of the four phases M (mitosis), G_1, S (DNA replication) and G_2, in embryonic cells it essentially reduces to a rapid succession of mitoses and interphases; the latter oscillations have also been observed in cell extracts.

Experimental Aspects of the Mitotic Oscillator. Central to the mechanism of the mitotic oscillator is the interaction of the cdc2 protein kinase with cyclin, a protein whose name reflects its periodic variation during the cell cycle. There exists a family of different cyclins, among which cyclin B is associated with the G_2/M transition. The accumulation of cyclin B prior to mitosis triggers the activation of cdc2 kinase with which it forms a complex referred to as maturation (or mitosis) promoting factor (MPF) in amphibian eggs (Murray and Hunt 1993). The simplest form of mitotic trigger mechanism is found in the early stages of amphibian development: in such rapidly dividing embryonic cells, where the period of the mitotic oscillator is of the order of 30 min, the sole accumulation of cyclin suffices to drive the cell cycle (Murray and Kirschner 1989a).

In yeast and somatic cells, the cell cycle is subjected to additional controls, e.g. before the initiation of DNA replication where the checkpoint has been designated as "start" (Cross et al. 1989; Hartwell and Weinert 1989). Here also, however, the transition between the G_2 and M phases is brought about by the periodic activation of cdc2 kinase, while the transition between the G_1 and S phases is controlled by G_1-specific cyclins and, in higher eukaryotes, by another protein kinase, cdk2, closely related to cdc2. Thus the G_1-S transition appears to be controlled by a second oscillator involving G_1-specific cyclins and cdk2 kinase (Dulic et al. 1992). In the following, we focus on the simpler situation that prevails in the early stages of amphibian embryogenesis, although it is likely that the key features of the mitotic control system in embryonic cells remain at the core of more complex cell cycle mechanisms (Cross et al. 1989; Murray and Kirschner 1989a, b; Murray and Hunt 1993).

Advances made over the last five years have permitted the characterization of the biochemical mechanism by which cdc2 kinase becomes activated (see Draetta 1990). Such a process primarily relies on the dephosphorylation of cdc2 kinase on a tyrosine residue located in the ATP-binding domain. It appears, however, that cdc2 kinase is fully active only when a threonine residue of the enzyme is phosphorylated.

Because of the primary role of cyclin in the activation of cdc2 kinase, cyclin degradation plays an essential role in the control of mitosis. Important in this respect is the observation that cdc2 kinase activates cyclin degradation; the mechanism of that activation appears to rely on a phosphorylation cascade containing at least one phosphorylation-dephosphorylation cycle, the product of which is the active form of a cyclin protease (Félix et al. 1990). According to Murray and Kirschner (1989b) and to Félix et al. (1990), the activation of cdc2 kinase by cyclin and the subsequent activation of cyclin protease by cdc2 kinase constitute a negative feedback loop on which a minimal cell cycle oscillator might be based.

A Minimal Cascade Model For the Mitotic Oscillator. Shown in Fig.4.3 is the minimal cascade model considered for the mitotic oscillator (Goldbeter 1991, 1995a). The model is based on the activation of cdc2 kinase from the inactive, tyrosine-phosphorylated form M^+ into the active, dephosphorylated form M (Draetta 1990). This reaction is catalyzed by the cdc25 phosphatase; the reverse reaction is catalyzed by the wee1 kinase (Murray and Hunt 1993). Driving the cascade (Murray and Kirschner 1989a) is the accumulation of the constantly synthesized cyclin, which brings about the activation of cdc2 through dephosphorylation by cdc25.

At the core of the oscillatory mechanism is the negative feedback loop due to the fact that cyclin promotes the activation of cdc2 while the latter enzyme triggers the activation of a protease that specifically degrades cyclin (Murray and Kirschner 1989b; Félix et al. 1990; Murray and Hunt 1993). In this minimal model, the transition from the inactive form X^+ to the active form X of the cyclin protease is catalyzed by cdc2 kinase, while the reverse reaction is catalyzed by a phosphatase.

The three variables of the model are the concentration of cyclin and the fractions of active cdc2 kinase and active cyclin protease. Sustained oscillations of the limit cycle type readily arise in this model as a result of negative feedback and of the time delays which arise naturally from the successive cycles of phosphorylation-dephosphorylation. This and other, more detailed models (Novak and Tyson 1993a, b) provide a molecular basis for the biochemical oscillator that governs the periodic onset of mitosis. That mitosis might be driven by a limit cycle oscillator had been considered for long (see Kauffman and Wille 1975; Winfree 1980), but the biochemical nature of the oscillator was unknown at the time.

As mentioned above, negative feedback arises in the cascade from the fact that cyclin promotes the activation of cdc2 kinase while the latter enzyme triggers the destruction of cyclin, through activation of a cyclin protease. If the mechanism of oscillations relies on this negative feedback, one would expect that raising the

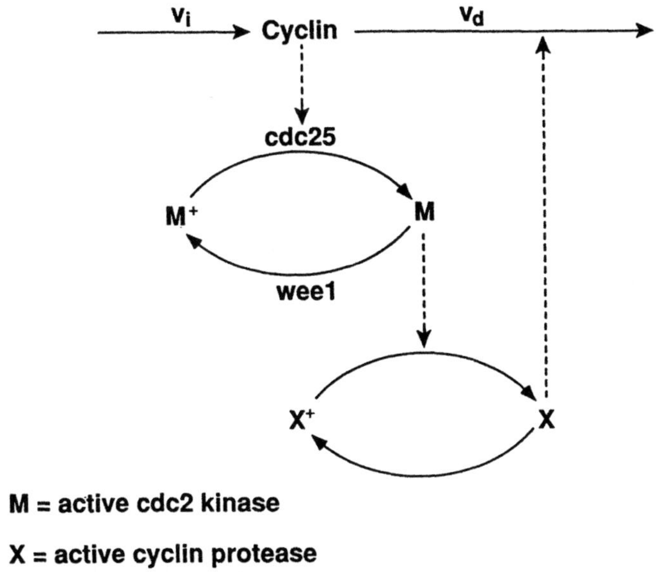

Fig.4.3 Scheme of the model for the mitotic oscillator in embryonic cells, based on the phosphorylation-dephosphorylation cascade involving cyclin and cdc2 kinase (Goldbeter 1991, 1995a)

rate of nonspecific cyclin degradation, measured in the model by the rate constant k_d (not shown in Fig.4.3), would bypass the feedback loop and would thereby attenuate or even suppress its destabilizing role. This is precisely what is shown by stability diagrams established for different values of parameter k_d (Romond et al 1994). The instability domain is indeed widest when this constant is nil. As the rate of nonspecific cyclin degradation rises, the domain of oscillations progressively shrinks and eventually vanishes when k_d approaches a sufficiently large value at which the nonspecific cyclin degradation rate ceases to be negligible with respect to the rate of specific degradation by protease X.

The phosphatase cdc25 is itself activated through phosphorylation and inactivated through dephosphorylation. Recent experiments showed that cdc2 kinase promotes its own activation by activating cdc25 through phosphorylation (Hoffman et al. 1993). Such autocatalysis plays a prominent role in several models proposed for the mitotic oscillator (Novak and Tyson 1993a, b). Incorporation of this additional phosphorylation-dephosphorylation cycle into the minimal cascade model of Fig.4.3 (Goldbeter 1993, 1995a) allows one to assess the contribution of autocatalysis to the instability mechanism giving rise to oscillations.

The analysis of such extended models indicates (Romond et al. 1994) that indirect positive feedback by cdc2 kinase, mediated through its activation of the cdc25 phosphatase, enlarges the domain of sustained oscillations in parameter space. What the basic cascade model shows, however, is that oscillations may arise solely as a result of negative feedback. Moreover, increasing the number of phosphorylation-dephosphorylation cycles in the cascade also contributes to enlarge the domain of sustained oscillations in parameter space. This is because each phosphorylation-dephosphorylation cycle on the way that leads from cyclin activation of cdc2 kinase to cyclin degradation by the cyclin protease introduces a time delay in the cascade. Time delays are known to favor the occurrence of sustained biochemical oscillations (see for example Landahl 1969).

Whereas the models show how the cyclin-driven cascade of phosphorylation-dephosphorylation cycles leading to cdc2 kinase activation can produce oscillations of about 30 min period, as observed in embryonic cells, the duration of the cell cycle in yeast and somatic cells is much longer. This lengthening of the period appears to result from the multiplicity of controls (or checkpoints) which behave as multiple brakes that slow down the biochemical oscillators which control the sequential onset of DNA replication and of mitosis in these cells.

4.3 Circadian Oscillations of the *period* Protein (PER) in Drosophila

As further emphasized elsewhere in this volume, circadian rhythms are ubiquitous as they occur in unicellular organisms (for review see Edmunds 1988) including algae such as Gonyaulax, Euglena, and Acetabularia, or fungi such as Neurospora, and in plants, insects such as the fly Drosophila (see below), molluscs, and mammals including rodents and humans. Some of the most remarkable advances in elucidating the molecular basis of circadian rhythms have been made in mutants of the fly Drosophila (Konopka 1979; Hall and Rosbash 1988; Baylies et al. 1993; Dunlap 1993). Two types of circadian rhythm are known in that organism. The first affects the rest-activity cycle, and the second underlies the daily eclosion peaks of pupae. Both rhythms persist in constant light or temperature (Pittendrigh 1960). After recalling the salient experimental results on circadian rhythms in Drosophila, a theoretical model recently proposed (Goldbeter 1995b) for circadian oscillations of the PER protein in this organism will briefly be presented.

4.3.1 Experimental Aspects of Circadian Oscillations of the *period* Protein (PER) in Drosophila

The classic work of Konopka and Benzer (1971) yielded Drosophila melanogaster flies altered in their circadian system, owing to mutations in a single gene that was called *per* (for *period*). Four phenotypes were characterized: the wild type (per^{+}) has a free-running period of activity and eclosion close to 24 h; short-period mutants (per^{S}) have a period close to 19 h; in long-period mutants (per^{L}), the periodicity increases up to 29 h; finally, arrhythmic mutants (per^{0})

have lost the circadian pattern of eclosion or activity (Konopka and Benzer 1971; Konopka 1979).

Interestingly, whereas the wild type exhibits the property of temperature compensation that is common to all circadian rhythms (Pittendrigh 1960), the mutants *per*^{l} and *per*^{S} have lost this property (Konopka et al. 1989); in contrast to the wild type, the period of their activity rhythm respectively increases and decreases with temperature. A molecular explanation for the temperature compensation of circadian oscillations in Drosophila has recently been proposed (Huang et al. 1995).

Surprisingly, the mutations in the *per* gene affect not only the circadian clock but also a high-frequency rhythm in Drosophila. Besides the circadian periodicity in eclosion and activity, Drosophila displays a rhythm of about 1 min period associated with the courtship song of the male. Kyriacou and Hall (1980) found that mutations of the *per* gene affect not only the period or even the existence of circadian rhythms, but also the much shorter-period rhythm of the courtship song. Moreover, the effect of the mutations keeps the same trend for both rhythms. A remarkable property of the *per* gene product therefore is that it affects 24 h rhythms as well as a 1 min rhythm.

A major cue to the role of the *per* locus in these rhythms has come from the molecular characterization of the gene and of its protein product (PER). These studies (Baylies et al. 1987; Yu et al. 1987) showed that the *per* gene codes for a protein of about 1220 amino acid residues (a certain polymorphism in length exists for PER in different Drosophila strains; see Baylies et al. 1993). Molecular mapping of the mutations furthermore indicated that the short- and long-period phenotypes *per*^{S} *and* *per*^{l} both result from the substitution of a single amino acid (Baylies et al. 1987; Yu et al. 1987). The arrhythmic phenotype *per*^{0} also arises from a single mutation, which results in a truncated protein (Baylies et al. 1987; Yu et al. 1987). This altered protein presumably has lost its function, so that flies carrying the mutation become arrhythmic.

A key question pertains to the function of PER in the mechanism of the circadian rhythms. Recent studies showed that PER possesses a dimerization domain that is common to several transcription factors (Huang et al. 1993). Together with the mainly nuclear localization of PER in the brain of adult flies (Liu et al. 1992), which turns out to be required for circadian rhythmicity (Baylies et al. 1993; Sehgal et al. 1994; Vosshall et al. 1994), this suggests that PER behaves as a regulator of transcription, possibly by interacting with transcription factors (Takahashi, 1992; Huang et al. 1993, 1995), since it does not seem to possess itself a DNA-binding domain or DNA-binding activity (Baylies et al. 1993; Huang et al. 1993). In such a manner the circadian rhythm in PER could direct the circadian variation of other proteins (Hardin et al. 1992; Huang et al. 1993), through the control of circadian-clock responsive elements (Takahashi, 1993).

A breakthrough for the mechanism of circadian rhythms in Drosophila was the finding (Hardin et al. 1990, 1992) that *per* mRNA is produced in a circadian manner. This periodic variation is accompanied by a circadian rhythm in the degree of abundance of PER; the peak in *per* mRNA precedes the peak in PER by 4-8 h (Zerr et al. 1990; Zeng et al. 1994). On the basis of this observation, Hardin et al. (1990, 1992) suggested that the Drosophila circadian rhythm results from a negative feedback exerted by the PER protein - or by some related product - on the synthesis of the *per* mRNA or on its translation; PER may act by binding to and removing a transcription activator (Zeng et al. 1994). Post-translational modification of PER is also involved in the mechanism of circadian oscillations. Thus, a recent study showed that PER is multiply phosphorylated (Edery et al. 1994). The authors suggested that the phosphorylation status of PER plays a role in the circadian oscillatory mechanism, perhaps by controlling the nuclear localization of PER and/or its degradation.

Overexpression of PER in the Drosophila eyes represses *per* transcription and suppresses circadian rhythmicity in these cells, without affecting circadian oscillations in other *per*-expressing cells in the brain, or the circadian rhythm in locomotor activity. This work shows that the action of PER on transcription is intracellular, and suggests that "each *per*-expressing cell contains an autonomous oscillator of which the *per* feedback loop is a component" (Zeng et al. 1994). Such a mechanism, based on negative autoregulation of transcription, has also been found in Neurospora (Aronson et al. 1994), and may represent a unified mechanism for the generation of circadian rhythmicity in a wide variety of experimental systems (Takahashi, 1992; Sassone-Corsi, 1994), in much the same way as the regulation of cdc2 kinase controls mitotic oscillations in different eukaryotic organisms (see Section 3.3).

4.3.2 Minimal Model for Circadian Oscillations of the Drosophila PER Protein

A theoretical model has recently been proposed for circadian oscillations in the Drosophila PER protein and its mRNA (Goldbeter, 1995b). This model is based on multiple phosphorylation of PER and on the inhibition of *per* transcription by a phosphorylated form of the protein. It can be viewed as a minimal model because it takes into account a limited number of phosphorylated residues. Besides accounting for the circadian rhythms in mRNA and protein level, the model shows how variations in parameters such as the rate of degradation of PER or the rate of its translocation into the nucleus may change the period of the oscillations, or even suppress rhythmic behavior. These results provide insights into ways by which modifications affecting the stability of PER may give rise to the per^S, per^L, or per^0 mutant phenotypes described above.

An alternative, more detailed model for circadian PER oscillations in Drosophila has been developed independently by Abbott et al. (1995) (M. Rosbash, personal communication; the results of that study were first presented in May 1994 at the 4th conference of the Society for Research on Biological Rhythms, Amelia Island, Florida). That model, which is also based on the negative feedback exerted by PER on *per* transcription, takes into account a larger number of phosphorylated residues and focuses on the role of PER phosphorylation in delaying the entry of the protein into the nucleus (Curtin et al. 1995). According to Curtin et al. (1995), this delay strengthens the capability of the negative feedback to produce robust circadian cycling in *per* expression.

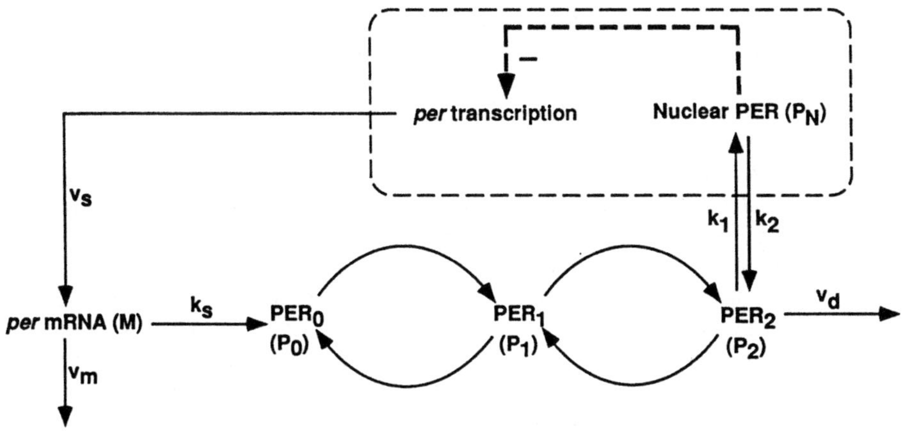

Fig.4.4 Scheme of the model for circadian oscillations of the PER protein in Drosophila, based on multiple phosphorylation of PER and on repression of *per* mRNA synthesis by a phosphorylated form of the protein (see text for details) (Goldbeter 1995a,b)

The minimal model for the circadian variation in PER is schematized in Fig.4.4. We assume that *per* mRNA, whose cytosolic concentration is denoted by M, is synthesized in the nucleus and rapidly transfers to the cytosol, where it accumulates at a maximum rate v_s; there it is degraded enzymically, in a Michaelian manner, at a maximum rate v_m. The rate of synthesis of PER, proportional to M, is characterized by an apparent first-order rate constant k_s. To take into account the fact that PER is multiply phosphorylated (Edery et al. 1994), and to keep the model as simple as possible (the precise number of phosphorylated residues is still unknown), only three states of the protein are considered: unphosphorylated (P_0), monophosphorylated (P_1) and bisphosphorylated (P_2). The model could readily be extended to include a larger number of phosphorylated residues; such an extension would, however, unnecessarily complicate the model without altering significantly its dynamic behavior. The model schematized in Fig.4.4 can be viewed as minimal with respect to the more detailed model considered by Abbott et al. (1995).

The role of PER phosphorylation is still unclear. It has been suggested that phosphorylation may control nuclear localization and/or degradation of PER (Edery et al. 1994). Here we assume that the fully phosphorylated form (P_2) is degraded enzymatically, at a maximum rate v_d, and also transported into the nucleus, at a rate characterized by the apparent first-order rate constant k_1. Transport of the nuclear, bisphosphorylated form of PER (P_N) into the cytosol is characterized by the apparent first-order rate constant k_2. The assumption that only the fully phosphorylated form of PER enters the nucleus holds with recent observations on delayed nuclear entry brought about by PER phosphorylation (Curtin et al. 1995). In the model, degradation of PER could also be directed at the nuclear, bisphosphorylated form of the protein, or at the cytosolic, unphosphorylated or monophosphorylated forms, which could also be transported into the nucleus. Probably, such changes would produce only minor modifications in dynamic behavior, although delaying PER entry into the nucleus and degradation until the protein is fully phosphorylated - as considered here - should favor the occurrence of sustained oscillations.

Crucial to the mechanism of oscillations in the model is the negative feedback exerted by nuclear PER on the production of *per* mRNA. This negative feedback is described by an equation of the Hill type in which n denotes the degree of cooperativity, and K_I the threshold constant for repression. To simplify the model, we consider that P_N behaves directly as a repressor; activation of a repressor upon binding of P_N would not significantly alter the results.

The time evolution of the model is governed by five kinetic equations for the variables M, P_0, P_1, P_2 and P_N (Goldbeter 1995a, b). The total (nonconserved) quantity of PER, P_t, is given by the relation $P_t = P_0 + P_1 + P_2 + P_N$.

Numerical integration shows that in a large domain of parameter values, instead of evolving toward a stable steady state, the system reaches a regime of sustained, periodic oscillations. Shown in Fig.4.5 is the temporal variation in the amount of *per* mRNA (M), together with the variation in the level of nuclear PER (P_N). Also shown are the periodic variation in the total amount of PER (P_t) and the concomitant changes in the amounts of unphosphorylated (P_0) and phosphorylated, cytosolic (P_1 and P_2) forms of PER.

As in the cases of ultradian oscillations discussed in Section 4.2, sustained oscillations in PER correspond to the evolution toward a limit cycle in the phase plane. This is demonstrated by plotting the level of *per* mRNA, M, as a function of the total amount of PER protein, P_t, for two different initial conditions, one located inside and the other outside the limit cycle (Goldbeter 1995a, b). In each case the system evolves toward the same, unique closed curve. Sustained oscillations in PER and *per* mRNA are thus characterized by a unique amplitude and frequency for a given set of parameter values, regardless of initial conditions. The limit cycle nature of circadian rhythms has been conjectured for long (Pavlidis 1973; Winfree 1980). Limit cycle models of an abstract mathematical nature or borrowed from physical sciences, e.g. the Van der Pol oscillator, have often been used to study properties of circadian rhythms (Wever 1972; Pavlidis 1973; Kronauer et al. 1982; Lakin-Thomas et al. 1991). The model for the periodic variation in PER and *per* mRNA provides a molecular basis for circadian oscillations of the limit cycle type.

The predictions of the model as to the period of the rhythm are of a qualitative rather than quantitative nature. Parameter values have been chosen so as to yield a period of the order of 24 h in Fig.4.5, given that detailed information is still lacking on the kinetics of the various steps considered in the model (see, e.g. Zeng et al. 1994). What the data of Fig.4.5 show, however, is that if the period is close to 24 h, then the phase shift between the peak of *per* mRNA and the peak of nuclear PER is of the order of 7 h. Such a result holds with the observation that the maximum in nuclear PER protein follows the peak in *per* mRNA by 4-8 h (Edery et al. 1994; Zeng et al. 1994).

Given the primary role of the negative feedback exerted by PER, one of the main parameters affecting oscillations in the model is v_d, which measures the

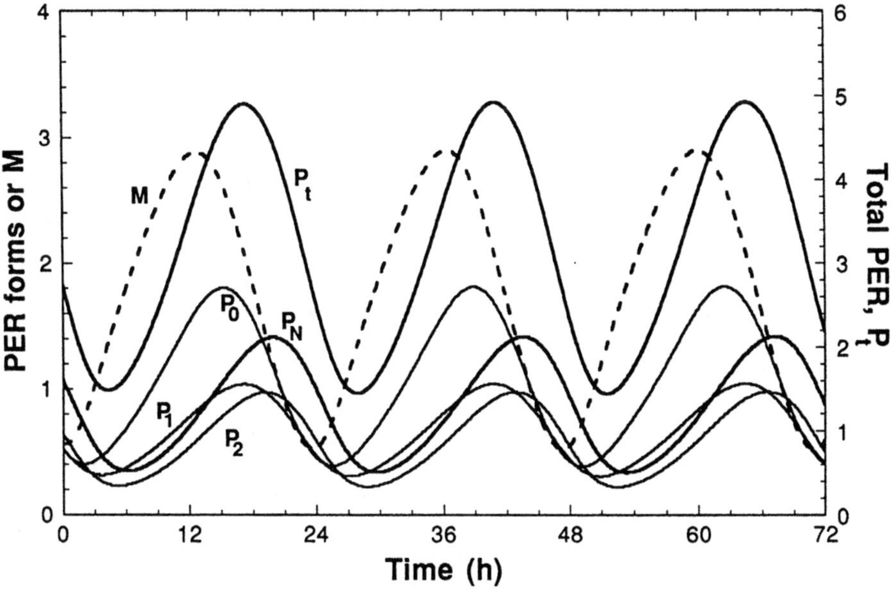

Fig.4.5 Typical oscillations generated by the model for circadian oscillations of the PER protein in Drosophila (see Fig.4.4 for a scheme of the model). Shown as a function of time are the concentrations of *per* mRNA (M) and of the various cytosolic, unphosphorylated (P_0) and phosphorylated (P_1, P_2) forms of PER, together with the nuclear, phosphorylated form (P_N) and the total PER concentration (P_t). The kinetic equations and parameter values used to generate the curves by numerical integration can be found in (Goldbeter 1995a,b)

maximum rate at which PER is degraded. Numerical simulations show that sustained oscillations occur in a window bounded by two critical values of this parameter (Goldbeter 1995a, b). In the interval of v_d values, the period of the oscillations increases typically from a value close to 19.3 h up to a value close to 64 h, under the conditions of Fig.4.5; for a larger value of parameter k_1, however, the interval of periods changes to 15-55 h as a function of v_d. The fact that the period of PER oscillations increases with the rate of PER degradation is due to the longer time required to reach the threshold beyond which PER significantly represses the transcription of its gene. The lengthening of the period is also due to the fact that the amplitude of the oscillations in M - and, consequently, in the various forms of PER - rises with v_d, since *per* mRNA has more time to accumulate before transcription shuts off.

The range of period found in the window of v_d values includes the range found in rhythmic *per* mutants of Drosophila. Thus, the short- and long-period mutants, per^S and per^L, have a period of 19 h and 28 h, respectively (Konopka and Benzer 1971). The ultrashort *per* mutant recently described (Konopka et al. 1994), which has a period close to 16 h, would also fall in the 15-55 h interval mentioned above. Longer periods, of the order of 50-60 h, such as those obtained in the model at high values of v_d in the situation considered, have not been observed in *per* mutants. If they exist, mutants possessing such a long-period rhythm might perhaps be taken as lacking any sign of periodic behavior.

If the period changes by a factor of about 3 as a function of v_d, it is reasonable to suppose that a similar variation can also be observed for other parameters of the model. Instead of the 15-55 h range, the close to threefold change in period might well cover, for example, the 8-27 h range. Then, upon changing the control parameter, the circadian oscillator could transform into an ultradian rhythm of 8 h period. Mutations leading to such transitions from circadian to ultradian rhythms have sometimes been observed (see Dowse and Ringo 1993). Nothing, in principle, precludes an oscillatory mechanism to produce circadian as well as ultradian periods, in different conditions, i.e. for different sets of parameter values.

The results obtained for the effect of parameter v_d suggest an explanation for the *per^s* and *per^l* phenotypes (Goldbeter 1995a, b). These results show that mutations that would affect the structure of PER so as to hinder or enhance its enzymatic degradation would result in a shortening or lengthening of the period, respectively. Rhythmic behavior may disappear altogether if the effect of mutations is such that the corresponding parameter (here, the maximum rate of PER degradation) quits the window in which oscillations occur. This could well happen for some arrhythmic *per* mutants.

The above explanation of *per* mutants in terms of variable PER degradation holds with the results of experiments in which the amount of *per* transcript was varied (Baylies et al. 1987). Baylies et al. concluded from that work that one possible interpretation of their results would be that "in *per^s* and *per^l*, amino-acid substitutions respectively increase or decrease the stability of the *per* protein"; a similar suggestion was made by Coté and Brody (1986). The variation in PER degradation rate may, however, not be the sole - or even the main - factor responsible for the alteration in period seen in *per* mutants. Thus, recent observations show that the delay in nuclear entry is normal for the *per^s* mutant, but increases in the mutant *per^l* (Curtin et al. 1995). Changes in the phosphorylation status of the protein may underlie these differences and may result, at the same time, in changed PER degradation rates.

Another important parameter that may control the oscillations is the apparent first-order rate constant k_1 that measures the transport of PER into the nucleus. Here again, there is again a range of k_1 values that produce sustained oscillations; the period of oscillations decreases as k_1 increases. The dependence of the period on parameter k_1 can be understood as for parameter v_d: when transport of P_2 into the nucleus is slower, P_N accumulates at a lower rate so that it takes more time for P_N to reach the threshold for repression of *per* transcription. In terms of parameter k_1, the *per^s* and *per^l* phenotypes might thus be respectively associated with enhanced and decreased rates of transport of PER into the nucleus.

Periodic behavior disappears if the transport rate into the nucleus decreases below the lower critical value. Such a situation could correspond to the loss of circadian rhythmicity in the *timeless* (*tim*) mutant in Drosophila (Sehgal et al. 1994), for which nuclear localization of PER appears to be blocked (Vosshall et al. 1994). A key role for transport of PER into the nucleus is consistent with the view that the negative feedback exerted by PER on transcription is at the core of the mechanism of circadian rhythmicity.

Although based on the evidence that has accumulated in recent years on the role of the *per* gene in the circadian clock of Drosophila, the model presented above and the more detailed one proposed by Abbott et al. (1995) might also pertain to circadian oscillations in other organisms. One of the most conspicuous features of circadian rhythms, encompassed by these models, is the role played by protein synthesis in the oscillatory mechanism. Such a role is attested by the observation, repeatedly made in different experimental systems, that inhibitors of protein synthesis, and also inhibitors of transcription, alter the period of circadian oscillations or produce phase shifts in these rhythms (for review see Takahashi et al. 1993).

Protein synthesis together with the negative control exerted by the PER protein on the synthesis of its mRNA (Hardin et al. 1990, 1992; Zeng et al. 1994) and the delay introduced in that negative feedback loop by PER phosphorylation (Curtin et al. 1995) play a key role in the mechanism proposed for circadian oscillations in Drosophila. A similar negative feedback on transcription has also been observed for the *frq* gene in Neurospora (Aronson et al. 1994) and could well prove to be a general feature of the mechanism of circadian rhythmicity in unicellular as well as multicellular organisms (Takahashi 1992, 1993; Sassone-Corsi 1994). The existence of a unified mechanism for the circadian clock - encompassing mRNA and protein synthesis, as well as autoregulatory feedback by a key protein behaving as a regulator of transcription - would parallel the situation encountered for the mitotic oscillator, which turns out to possess universal features among all eukaryotic organisms.

4.4 Concluding Remarks

Early after the discovery of glycolytic oscillations, it was suggested that circadian rhythms may originate from metabolic oscillations, which occur in oscillatory enzyme reactions with a period of several minutes, if the enzyme activities are sufficiently slowed down. The period of glycolytic oscillations was accordingly extended, but only up to the order of 1 h, by decreasing the temperature (Chance et al. 1964). Models for biochemical oscillations were used to investigate properties of circadian rhythms such as temperature compensation (Pavlidis and Kauzmann 1969). The relationship between ultradian biochemical oscillations and circadian rhythms was explored here in further detail. We examined three of the best-known biochemical oscillators and summarized what is known of their underlying mechanism from an experimental and theoretical point of view. We then turned to circadian oscillations in the PER protein in Drosophila and presented a minimal model for the generation of this circadian biochemical rhythm. Nothing fundamental, besides the period and the nature of the molecules involved, distinguishes the mechanisms of these ultradian or circadian biochemical oscillators. The models indicate that in each case sustained oscillations are of the limit cycle type. The oscillatory mechanisms which underlie these different biochemical rhythms are further compared below.

Both glycolytic and cAMP oscillations rely on positive feedback and occur with a period of the order of 5 to 10 min. The mitotic oscillator, which takes its simplest form in amphibian embryonic cells, relies on a mixture of negative and positive feedback; its period is longer, of the order of 30 min in embryonic cells. The cell cycle possesses a much longer period in somatic cells, as the underlying oscillators, which govern the successive G_1/S and G_2/M transitions, are then subjected to additional controls which represent as many brakes on their cycling.

The period of glycolytic oscillations is primarily determined by the interval between two peaks of product generated by the autocatalytic PFK reaction. This interval is itself due to the replenishment of substrate, which is injected at a constant rate, and, to a lesser degree, to the much more rapid return of the oscillatory enzyme to its active conformation via the allosteric transition. In the case of cAMP oscillations in Dictyostelium, which involve the extracellular and intracellular compartments separated by the plasma membrane, the period is mainly determined by the return of the cAMP membrane receptor to its active conformation. It is here that we encounter for the first time a role for phosphorylation-dephosphorylation in the mechanism of a biochemical rhythm: the receptor is indeed desensitized through cAMP-induced phosphorylation by a protein kinase, and the time required for resensitization is set by the phosphatase that brings the receptor back to its active conformation.

Regulation by phosphorylation-dephosphorylation is also at the core of the biochemical mechanism underlying the mitotic oscillator that drives the embryonic cell cycle. The oscillations rely on a cascade of phosphorylation-dephosphorylation cycles driven by the synthesis of cyclin whose accumulation brings about the activation of cdc2 kinase; the latter in turn induces the activation of a cyclin protease that brings an end to cdc2 kinase activation. Besides this negative feedback, the fact that the phosphatase cdc25, which activates cdc2 kinase, is itself activated by cdc2 kinase introduces a positive feedback loop. It appears, however, that although its role in the oscillatory mechanism may be important, such indirect autocatalysis does not represent a prerequisite for rhythmic behavior since oscillations may arise solely from cdc2-induced cyclin degradation (Goldbeter 1991, 1995a). The longer period of the mitotic oscillator in embryonic cells (as compared to glycolytic or cAMP oscillations) results from the time required by cyclin to accumulate up to the threshold at which cdc2 kinase is activated and from the delays due to the successive phosphorylation-dephosphorylation cycles acting on cdc2 kinase and on the cyclin protease.

The minimal model for circadian oscillations of the PER protein in Drosophila (Goldbeter 1995 a, b) and the more detailed model proposed by Abbott et al. (1995) also involve protein synthesis, negative feedback, and regulation through phosphorylation-dephosphorylation. Here, however, the change in period is more significant as the period increases by more than one order of magnitude when going from the embryonic cell cycle clock to circadian oscillations. A first difference between the models schematized in Figs.2 and 3 is that the negative feedback by PER operates at the level of transcription; this autoregulatory control of the synthesis of PER is accompanied by a succession of phosphorylation-dephosphorylation cycles, which presumably prepare the protein both for transit into the nucleus and enzymatic degradation. The negative feedback exerted by PER on *per* mRNA synthesis is therefore delayed by these phosphorylation-dephosphorylation cycles (Curtin et al. 1995) and by the shuttling of the protein and its mRNA between the cytosol and the nucleus. The combination of these delays in biochemical processing and transport across the nuclear membrane conveys a circadian period upon the oscillations produced by the negative feedback loop.

The analysis of the minimal model for PER oscillations nevertheless shows that the period of oscillations may often vary by a factor of 3 or more as a function of a control parameter such as the maximum rate of PER degradation. This indicates that the same biochemical control mechanism may, in principle, produce oscillations of circadian or ultradian period, depending on the values of the parameters. These values may well be changed by mutations, which suggests an explanation for the observation that circadian rhythms sometimes give way to ultradian oscillations. An alternative explanation of the latter observation relies on the consideration of multiple oscillators whose output, prior to desynchronization, possesses a circadian nature (Dowse and Ringo 1993).

The minimal model proposed for circadian oscillations in PER protein and *per* mRNA is closely related to the model proposed some 30 years ago by Goodwin (1965), who discussed the conditions in which a protein repressing the transcription of its gene can produce sustained oscillations in the levels of that protein and its mRNA. Because they were not sufficiently nonlinear, the equations originally proposed by Goodwin did not give rise to limit cycle oscillations. Equations of that type were investigated for limit cycle behavior in further detail by Griffith (1968), and used explicitly for circadian rhythms, to determine phase response curves with respect to transient perturbations (Drescher et al. 1982). Similar equations were also studied for the related problem of how sustained oscillations occur in a metabolic chain of enzyme reactions regulated by end-product inhibition (Morales and McKay 1967; Walter 1970; Hunding 1974; Rapp 1975; Tyson and Othmer 1978). An experimental system studied by means of closely related models (Bliss et al. 1982) is the oscillatory synthesis of tryptophan in Escherichia coli, in which negative control by the end product is exerted both at the genetic and enzymatic levels. Oscillations with a period of the order of 1 h have been observed experimentally in that system, in a mutant characterized by reduced feedback inhibition at the enzyme level (Bliss et al. 1982).

The main result coming from these theoretical studies was that the occurrence of sustained oscillations is favored both by enlarging the length of the reaction chain that leads from the regulated step to the end product, and by increasing the degree of cooperativity of the negative feedback process. These results bear on the model for PER oscillations. Thus, the sequence of successive phosphorylations of the PER protein can be viewed as playing a role similar to that of increasing the number of intermediate steps in the chain of enzyme reactions regulated by end-product inhibition. A similar situation arises in the cascade model for the mitotic oscillator (see Section 2.3 and Romond et al. 1994). Incorporating more than two phosphorylation steps in the model for the circadian rhythm in PER accordingly enlarges the domain of sustained oscillations.

With regard to the role of nonlinear feedback control, the periodic behavior shown in Fig.4.5 occurs for a repression function characterized by a cooperativity coefficient of 4. A value of 2, or even 1, for that coefficient can also give rise to sustained oscillations; the domain of oscillations in parameter space, however, is then smaller than for $n = 4$. This shows that if the cooperativity of repression favors periodic behavior, multiple phosphorylation of PER, by introducing a series of delays, could substitute for (or reinforce the effect of) of such cooperativity in allowing for sustained oscillations. This conclusion supports the view (Curtin et al. 1995; Abbott et al. 1995) that, by gating PER entry into the nucleus, PER phosphorylation delays the negative feedback exerted by PER on *per* transcription, and thereby strengthens the capability of such feedback to produce robust circadian oscillations. The puzzle remains as to how PER may at the same time participate in the generation of both circadian and very rapid ultradian rhythmicity, as shown by the effect of *per* mutations on the 1 min rhythm characterizing the Drosophila courtship song (Kyriacou and Hall 1980).

Ultradian biochemical oscillations often rely on positive feedback, as illustrated here by the cases of glycolytic and cAMP oscillations. The mechanism of the mitotic oscillator appears to involve a combination of positive and negative feedback loops. Providing further evidence for the variety of mechanisms that can produce sustained oscillations in biological systems, studies in Drosophila and Neurospora, corroborated by the analysis of theoretical models, suggest that a negative feedback at the level of transcription is involved in the generation of circadian rhythms in these and other organisms.

Acknowledgments

This work was supported by grant 3.4588.93 from the Fonds de la Recherche Scientifique Médicale (FRSM, Belgium) and by the Programme "Actions de Recherche Concertée" (convention 94/99-180) launched by the Division of Science and Higher Education, French Community of Belgium.

References

Abbott L, Zeng H, Rosbash M (1995) A model of circadian rhythm generation in *Drosophila*. Submitted for publication.

Alcantara F, Monk M (1974) Signal propagation during aggregation in the slime mould *Dictyostelium discoideum*. J Gen Microbiol 85:321-334.

Aronson BD, Johnson KA, Loros JL, Dunlap JC (1994) Negative feedback defining a circadian clock: autoregulation of the clock gene *frequency*. Science 263:1578-1584.

Baylies MK, Bargiello TA, Jackson FR, Young MW (1987) Changes in abundance or structure of the *per* gene product can alter periodicity of the *Drosophila* clock. Nature 326:390-392.

Baylies MK, Weiner L, Vosshall LB, Saez L, Young MW (1993) Genetic, molecular, and cellular studies of the *per* locus and its products in *Drosophila melanogaster*. In: Young MW (ed) Molecular genetics of biological rhythms. Marcel Dekker, New York, pp 123-153.

Berridge MJ, Rapp PE (1979) A comparative survey of the function, mechanism and control of cellular oscillations. J Exp Biol 81:217-279.

Bliss RD, Painter PR, Marr AG (1982) Role of feedback inhibition in stabilizing the classical operon. J Theor Biol 97:177-193.

Boiteux A, Goldbeter A, Hess B (1975) Control of oscillating glycolysis of yeast by stochastic, periodic, and steady source of substrate: a model and experimental study. Proc Natl Acad Sci USA 72:3829-3833.

Bonner JT (1967) The cellular slime molds. Princeton University Press, Princeton, NJ.

Chance B, Schoener B, Elsaesser S (1964) Control of the waveform of oscillations of the reduced pyridine nucleotide level in a cell-free extract. Proc Natl Acad Sci USA 52:337 341.

Chance B, Pye K, Higgins J (1967) Waveform generation by enzymatic oscillators. IEEE Spectrum 4:79-86.

Chou HF, Berman N, Ipp E (1992) Oscillations of lactate released from islets of Langerhans: evidence for oscillatory glycolysis in b-cells. Am J Physiol 262:E800-E805.

Coté GG, Brody S (1986) Circadian rhythms in *Drosophila melanogaster*: analysis of period as a function of gene dosage at the *per* (period) locus. J Theor Biol 121:487-503.

Cross F, Roberts J, Weintraub H (1989) Simple and complex cell cycles. Annu Rev Cell Biol 5:341-395.

Curtin KD, Huang ZJ, Rosbash M (1995) Temporally regulated nuclear entry of the Drosophila period protein contributes to the circadian clock. Neuron 14:365-372.

Devreotes PN (1989) *Dictyostelium discoideum*: A model system for cell-cell interactions in development. Science 245:1054-1058.

Devreotes PN, Sherring JA (1985) Kinetics and concentration dependence of reversible cAMP-induced modification of the surface cAMP receptor in *Dictyostelium*. J Biol Chem 260:6378-6384.

Dowse HB, Ringo JM (1993) Is the circadian clock a "meta-oscillator"? Evidence from studies of ultradian rhythms in *Drosophila*. In: Young MW (ed) Molecular genetics of biological rhythms. Marcel Dekker, New York, pp 195-220.

Draetta G (1990) Cell cycle control in eukaryotes: Molecular mechanism of cdc2 activation. Trends Biochem Sci 15:378-383.

Drescher K, Cornelius G, Rensing L (1982) Phase response curves obtained by perturbing different variables of a 24 hr model oscillator based on translational control. J Theor Biol 94:345-353.

Dulic V, Lees E, Reed SI (1992) Association of human cyclin E with a periodic G_1-S phase protein kinase. Science 257:1958-1961.

Dunlap JC (1993) Genetic analysis of circadian clocks. Annu Rev Physiol 55:683-728.

Durston AJ (1974) Pacemaker activity during aggregation in *Dictyostelium discoideum*. Dev Biol 37:225-235.

Edery I, Zwiebel LJ, Dembinska ME, Rosbash M (1994) Temporal phosphorylation of the *Drosophila* period protein. Proc Natl Acad Sci USA 91:2260-2264.

Edmunds LN Jr (1988) Cellular and molecular bases of biological clocks. Models and mechanisms for circadian timekeeping. Springer, New York.

Félix MA, Labbé JC, Dorée M, Hunt T, Karsenti E (1990) Triggering of cyclin degradation in interphase extracts of amphibian eggs by cdc2 kinase. Nature 346:379-382.

Frenkel R (1968) Control of reduced diphosphopyridine nucleotide oscillations in beef heart extracts. I. Effect of modifiers of phosphofructokinase activity. Arch Biochem Biophys 125:151-156.

Gerisch G (1987) Cyclic AMP and other signals controlling cell development and differentiation in *Dictyostelium*. Annu Rev Biochem 56:853-879.

Gerisch G, Hess B (1974) Cyclic-AMP controlled oscillations in suspended *Dictyostelium* cells: their relation to morphogenetic cell interactions. Proc Natl Acad Sci USA 71:2118-2122.

Gerisch G, Malchow D, Roos W, Wick U (1979) Oscillations of cyclic nucleotide concentrations in relation to the excitability of *Dictyostelium* cells. J Exp Biol 81:33-47.

Gerisch G, Wick U (1975) Intracellular oscillations and release of cyclic AMP from *Dictyostelium* cells. Biochem Biophys Res Commun 65:364-370.

Ghosh AK, Chance B (1964) Oscillations of glycolytic intermediates in yeast cells. Biochem Biophys Res Commun 16:174-181.

Goldbeter A (1991) A minimal cascade model for the mitotic oscillator involving cyclin and cdc2 kinase. Proc Natl Acad Sci USA 88:9107-9111.

Goldbeter A (1993) Modeling the mitotic oscillator driving the cell division cycle. Comments Theor Biol 3:75-107.

Goldbeter A (1995a) Biochemical oscillations and cellular rhythms: the molecular bases of periodic and chaotic behaviour. Cambridge University Press, Cambridge, UK (in press).

Goldbeter A (1995b) A model for circadian oscillations in the *Drosophila period* protein (PER). Proc R Soc Lond B (in press).

Goldbeter A, Caplan SR (1976) Oscillatory enzymes. Annu Rev Biophys Bioeng 5:449-476.

Goldbeter A, Lefever R (1972) Dissipative structures for an allosteric model. Application to glycolytic oscillations. Biophys J 12:1302-1315.

Goldbeter A, Martiel JL (1988) Developmental control of a biological rhythm: the onset of cAMP oscillations in *Dictyostelium* cells. In: Markus M, Müller S, Nicolis G (eds) From chemical to biological organization. Springer, Berlin, pp 248-254.

Goldbeter A, Nicolis G (1976) An allosteric enzyme model with positive feedback applied to glycolytic oscillations. In: Snell F, Rosen R (eds) Progress in theoretical biology, vol 4. Academic Press, New York, pp 65-160.

Goldbeter A, Segel LA (1980) Control of developmental transitions in the cyclic AMP signaling system of *Dictyostelium discoideum*. Differentiation 17:127-135.

Goodwin BC (1965) Oscillatory behavior in enzymatic control processes. Adv Enzyme Regul 3:425-438.

Griffith JS (1968) Mathematics of cellular control processes. I. Negative feedback to one gene. J Theor Biol 20:202-208.

Gundersen RE, Johnson R, Lilly P, Pitt G, Pupillo M, Sun T, Vaughan R, Devreotes PN (1989) Reversible phosphorylation of G-protein-coupled receptors controls cAMP oscillations in *Dictyostelium*. In: Goldbeter A (ed) Cell to cell signalling: from experiments to theoretical models. Academic Press, London, pp 477-488.

Hall JC, Rosbash M (1988) Mutations and molecules influencing biological rhythms. Annu Rev Neurosci 11:373-393.

Hardin PE, Hall JC, Rosbash M (1990) Feedback of the *Drosophila period* gene product on circadian cycling of its messenger RNA levels. Nature 343:536-540.

Hardin PE, Hall JC, Rosbash M (1992) Circadian oscillations in period gene mRNA levels are transcriptionally regulated. Proc Natl Acad Sci USA 89:11711-11715.

Hartwell LH, Weinert TA (1989) Checkpoints: Controls that ensure the order of cell cycle events. Science 246:629-634.
Hess B, Boiteux A (1968) Control of glycolysis. In: Järnefelt J (ed) Regulatory functions of biological membranes. Elsevier, Amsterdam, pp 148-162.
Hess, B, Boiteux A (1971) Oscillatory phenomena in biochemistry. Annu Rev Biochem 40:237-258.
Hess B, Boiteux A, Krüger J (1969) Cooperation of glycolytic enzymes. Adv Enzyme Regul 7:149-167.
Higgins J (1964) A chemical mechanism for oscillation of glycolytic intermediates in yeast cells. Proc Natl Acad Sci USA 51:989-994.
Hoffmann I, Clarke PR, Marcote MJ, Karsenti E, Draetta G (1993) Phosphorylation and activation of human cdc25-C by cdc2-cyclin B and its involvement in the self-amplification of MPF at mitosis. EMBO J 12:53-63.
Huang ZJ, Edery I, Rosbash M (1993) PAS is a dimerization domain common to *Drosophila* Period and several transcription factors. Nature 364:259-262.
Huang ZJ, Curtin KD, Rosbash M (1995) PER protein interactions and temperature compensation of a circadian clock in *Drosophila*. Science 267:1169-1172.
Hunding A (1974) Limit cycles in enzyme systems with nonlinear negative feedback. Biophys Struct Mechan 1:47-54.
Kauffman S, Wille JJ (1975) The mitotic oscillator in *Physarum polycephalum*. J Theor Biol 55:47-93.
Konopka RJ (1979) Genetic dissection of the *Drosophila* circadian system. Fed Proc 38:2602-2605.
Konopka RJ, Benzer S (1971) Clock mutants of *Drosophila melanogaster*. Proc Natl Acad Sci USA 68:2112-2116.
Konopka RJ, Hamblen-Coyle MJ, Jamison CF, Hall JC (1994) An ultrashort clock mutation at the period locus of *Drosophila melanogaster* that reveals some new features of the fly's circadian system. J Biol Rhythms 9:189-216.
Konopka RJ, Pittendrigh CS, Orr D (1989) Reciprocal behaviour associated with altered homeostasis and photosensitivity of *Drosophila* clock mutants. J Neurogenet 6:1-10.
Kronauer RE, Czeisler CA, Pilato S, Moore-Ede MC, Weitzman ED (1982) Mathematical model of the human circadian system with two interacting oscillators. Am J Physiol 242:R3-R17.
Kyriacou CP, Hall JC (1980) Circadian rhythm mutations in *Drosophila melanogaster* affect the short-term fluctuations in the male's courtship song. Proc Natl Acad Sci USA 77:6729-6733.
Lakin-Thomas PL, Brody S, Coté GG (1991) Amplitude model for the effects of mutations and temperature on period and phase resetting of the *Neurospora* circadian oscillator. J Biol Rhythms 6:281-297.
Landahl HD (1969) Some conditions for sustained oscillations in biochemical chains with feedback inhibition. Bull Math Biophys 31:775-787.
Liu X, Zwiebel LJ, Hinton D, Benzer S, Hall JC, Rosbash M (1992) The *period* gene encodes a predominantly nuclear protein in adult *Drosophila*. J Neurosci 12:2735-2744.
Lloyd D, Stupfel M (1991) The occurrence and functions of ultradian rhythms. Biol Rev 66:275-299.
Martiel JL, Goldbeter A (1987) A model based on receptor desensitization for cyclic AMP signaling in *Dictyostelium* cells. Biophys J 52:807-828.
Morales M, McKay D (1967) Biochemical oscillations in "controlled" systems. Biophys J 7:621-625.
Murray AW, Hunt T (1993) The cell cycle: an introduction. Oxford University Press, Oxford.

Murray AW, Kirschner MW (1989a) Cyclin synthesis drives the early embryonic cell cycle. Nature 339:275-280.
Murray AW, Kirschner MW (1989b) Dominoes and clocks: The union of two views of the cell cycle. Science 246:614-621.
Novak B, Tyson JJ (1993a) Modeling the cell division cycle: M-phase trigger, oscillations, and size control. J Theor Biol 165:101-134.
Novak B, Tyson JJ (1993b) Numerical analysis of a comprehensive model of M-phase control in *Xenopus* oocytes extracts and intact embryos. J Cell Science 106:1153-1168.
Nurse P (1990) Universal control mechanism regulating onset of M-phase. Nature 344:503-508.
O'Rourke B, Ramza BM, Marban E (1994) Oscillations of membrane current and excitability driven by metabolic oscillations in heart cells. Science 265:962-966.
Pavlidis T (1973) Biological oscillators: their mathematical analysis. Academic Press, New York.
Pavlidis T, Kauzmann W (1969) Toward a quantitative biochemical model for circadian oscillators. Arch Biochem Biophys 132:338-348.
Pittendrigh C (1960) Circadian rhythms and the circadian organization of living systems. Cold Spring Harbor Symp Quant Biol XXV:159-184.
Pye EK (1969) Biochemical mechanisms underlying the metabolic oscillations in yeast. Can J Bot 47:271-285.
Pye EK, Chance B (1966) Sustained sinusoidal oscillations of reduced pyridine nucleotide in a cell-free extract of *S. carlsbergensis*. Proc Natl Acad Sci USA 55:888-894.
Rapp PE (1975) A theoretical investigation of a large class of biochemical oscillators. Math Biosci 25:165-188.
Romond PC, Guilmot JM, Goldbeter A (1994) The mitotic oscillator: Temporal self-organization in a phosphorylation-dephosphorylation enzymatic cascade. Ber Bunsenges Phys Chem 98:1152-1159.
Sassone-Corsi P (1994) Rhythmic transcription and autoregulatory loops: winding up the biological clock. Cell 78:361-364.
Sehgal A, Price JL, Man B, Young MW (1994) Loss of circadian behavioral rhythms and *per* RNA oscillations in the *Drosophila* mutant *timeless*. Science 263:1603-1606.
Sel'kov EE (1968) Self-oscillations in glycolysis. 1. A simple kinetic model. Eur. J. Biochem. 4:79-86.
Takahashi JS (1992) Circadian clock genes are ticking. Science 258:238-240.
Takahashi JS (1993) Circadian-clock regulation of gene expression. Curr Opin Genet Dev 3:301-309.
Takahashi JS, Kornhauser JM, Koumenis C, Eskin A (1993) Molecular approaches to understanding circadian oscillations. Annu Rev Physiol 55:729-753.
Tornheim K, Andrés V, Schulz V (1991) Modulation by citrate of glycolytic oscillations in skeletal muscle extracts. J Biol Chem 266:15675-15678.
Tyson JJ, Othmer HG (1978) The dynamics of feedback control circuits in biochemical pathways. In: Snell F, Rosen R (eds) Progress in Theoretical Biology, vol. 5. Academic Press, New York, pp 2-62.
Von Klitzing L, Betz A (1970) Metabolic control in flow systems. I. Sustained glycolytic oscillations in yeast suspensions under continuous substrate infusion. Arch Mikrobiol 71:220-225.
Vosshall LB, Price JL, Sehgal A, Saez L, Young MW (1994) Block in nuclear localization of *period* protein by a second clock mutation, *timeless*. Science 263:1606-1609.
Walter C (1970) The occurrence and the significance of limit cycle behavior in controlled biochemical systems. J Theor Biol 27:259-272.
Wever RA (1972) Virtual synchronization towards the limits of the range of entrainment. J Theor Biol 36:119-132.

Winfree AT (1972) Spiral waves of chemical activity. Science 175:634-636.
Winfree AT (1980) The geometry of biological time. Springer, New York (reprinted as Springer Study Edition, 1990, Springer, Berlin).
Yu Q, Jacquier AC, Citri Y, Hamblen M, Hall JC, Rosbash M (1987) Molecular mapping of point mutations in the *period* gene that stop or speed up biological clocks in *Drosophila melanogaster*. Proc Natl Acad Sci USA 84:784-788.
Zeng H, Hardin PE, Rosbash M (1994) Constitutive overexpression of the *Drosophila period* protein inhibits *period* mRNA cycling. EMBO J 13:3590-3598.
Zerr DM,Rosbash M, Hall JC, Siwicki KK (1990) Circadian fluctuations of *period* protein immunoreactivity in the CNS and the visual system of *Drosophila*. J Neurosci 10:2749-2762.

5 Cross-talk Between Clocks: Regulation of Cell Division Cycles by Circadian Oscillators[2]

Leland N. Edmunds, Jr.
Division of Biological Sciences, State University of New York at Stony Brook, Stony Brook, New York 11794-5215, USA

Summary

The achlorophyllous ZC mutant of the unicell Euglena gracilis exhibits a circadian rhythm of cell division in constant darkness (DD) . Mitosis occurs during a restricted part of the circadian cycle, corresponding to the dark intervals in LD:12,12.

This 'gating' phenomenon is thought to reflect an interaction between a circadian clock and the cyclin-p34^{cdc2} oscillator that has been demonstrated to drive the cell division cycle (CDC) in a variety of organisms. We are attempting to elucidate this signal transduction pathway. We have shown that division-phased cultures also exhibit bimodal circadian changes of cAMP level. Maximum cAMP levels occur at the beginning of light (CT 0-2) and at the onset of darkness (CT 12-14). These variations persist in DD and appear to be under the control of the circadian oscillator. Recently, we have discovered that adenylate cyclase and phosphodiesterase - the two enzymes whose dynamic balance determines the intracellular level of cAMP- also exhibit inverse, bimodal circadian changes in their own levels in both dividing and nondividing cultures. The level of cyclic GMP, a modulator of cAMP activity, also displays a persisting bimodal circadian variation.

---

[2] This review, based on a presentation at a workshop on 'The Molecular Mechanism of Circadian Rhythms,' held April 10-15, 1994 at Wohldenberg, Germany (see Edmunds 1994, 1995), summarizes recent work from my laboratory and has been supported in part by grants DCB-8901944 and -9105752 from the National Science Foundation. Particular credit is due to Drs. Isabelle A. Carré, Gangaram Mohabir and Jian Tong for the use of their published and unpublished results.

On the hypothesis that cAMP may be a coupling link between the circadian clock and the CDC, conditions that override the control of the cAMP level by the oscillator should also cause perturbations of the cell division rhythm. Indeed, we found that the addition of cAMP resulted in the temporary uncoupling of the CDC from the circadian timer. Pulses of cAMP given during the subjective day resulted in delays (up to 9 h) of the next synchronous division step; in contrast, mitosis was advanced when cAMP was administered in the middle of the subjective night. These $\pm \Delta\phi$s of the division steps reflected real changes in the rate of CDC progression. Measurement of cellular DNA content by flow cytometry indicated that cAMP pulses at CT 06-08 delayed progression through S phase (and perhaps also through mitosis) but at CT 18-20 accelerated the G_2 / M transition. The circadian oscillator was not perturbed by the cAMP pulse: the division rhythm soon returned to its original phase. Further, if the periodic cAMP signals generated by the oscillator were suppressed by the addition of the cAMP analog forskolin, an activator of adenylate cyclase that maintained cAMP at an abnormally high level, division rhythmicity was lost in DD, and even in LD:12,12. These findings indicate that although cAMP signals do not reset the circadian oscillator itself in Euglena (and thus, that cAMP is unlikely to represent a 'gear' of the clock), they do regulate CDC progression and may be sufficient to generate cell division rhythmicity.

How do these cAMP signals from the clock effect their regulation of the CDC? We have shown that the receptors for cAMP in Euglena are two types of cAMP-dependent kinase (cPKA and cPKB). Pharmacological studies using cAMP analogs suggest that cPKA mediates cAMP effects during the subjective day, and cPKB during the subjective night. We are now examining the regulation of cAMP-dependent kinases during the CDC (and during the circadian cycle), and are attempting to identify the targets that are selectively phosphorylated by these enzymes. Finally, we have begun to search for a homolog of p34 (the cdc 2 protein kinase gene product of fission yeast) in Euglena.

How does the circadian oscillator, acting through (or in conjunction with) the AC-cAMP-PDE system and the cAMP-dependent kinases (cPKA and cPKB) that we have identified, ultimately couple to the MPF clock regulating the CDC? Preliminary western blot analyses of protein extracts from dividing and stationary cultures of the ZC mutant in DD confirm the presence of a homolog of $p34^{cdc2}$, the catalytic subunit of MPF and suggest that its abundance throughout the CDC is invariant, decreasing to a much reduced level when cells cease dividing. Further, a relative shift in electrophoretic mobility of the monomer was observed to occur in a cycle-dependent manner: a more slowly moving band appeared at CT 04 and disappeared at CT 12. Such changes in mobility have been attributed to phosphorylation and dephosphorylation. These initial findings, if replicated, suggest that the activity of this polypeptide depends on posttranslational mechanisms, and, thus we have turned our attention to the regulatory subunit(s) of MPF. On the assumption that cyclin B is a likely candidate, we have begun western blot analyses using monoclonal antibodies and have found that, in contrast to $p34^{cdc2}$, the abundance of cyclin B in protein extracts varied quantitatively during both the circadian cycle and the CDC, being detected only at CT 20 and CT 24 (00). Further characterization of this key player in CDC regulation would relate to its state of phosphorylation.

5.1. Introduction

Among other mechanisms that have been proposed to control microbial cell division cycles (CDC), autonomous biological oscillators provide a possible means of integrating the disparate concepts of cell cycle transit, exit and arrest. An important goal is to determine the nature of the circadian clock that couples to the CDC under certain conditions and generates periodicity of division and other CDC 'landmarks' (as well as to underlie circadian rhythmicity in general). In this overview we are concerned with the means whereby the coupling between the circadian oscillator, on the one hand, and the cyclin–p34$^{\underline{cdc2/CDC28}}$ (MPF) clock that constitutes a universal cell-division-cycle 'engine,' on the other, is effected. Cyclic AMP (cAMP), which controls certain rate-limiting steps in the progression of the CDC in many cell types, may play such a coupling role, participating in the gating of CDC events to specific phases of the circadian cycle. Thus, we hope to elucidate the signal transduction pathway whereby information regarding period and phase flows from the circadian clock to the CDC and to determine the links (checkpoints) between the two oscillator systems.

To this end, we have intensively investigated circadian clock control of the CDC in the algal flagellate Euglena gracilis Klebs, a highly rhythmic system (Edmunds 1982; Edmunds and Halberg 1981) that has been well characterized physiologically and biochemically (Edmunds 1965b; Buetow 1968a,b, 1982). The formal properties of circadian clocks – in particular, entrainability, persistence, phase shiftability, and temperature compensation of the free-running period (τ) – have been found to characterize the circadian rhythm (CR) of cell division in both the wild type (Edmunds and Laval-Martin 1984) and achlorophyllous, photosynthesis-deficient ZC mutant (Carré et al. 1989a). We now summarize recent results from our laboratory and those of others that suggest a key role for cAMP (which exhibits bimodal circadian rhythmicity in Euglena) in circadian clock regulation of cell division cycles in this unicell and other cell types.

5.1.1 Regulation of the Cell Division Cycle

The cell division cycle of eukaryotes is a complex cascade of events that culminate in cell duplication (Cross et al. 1989). Rapid progress has been made within the past several years towards dissecting the regulatory network that provides precise coordination of these processes. A unifying view has emerged: the CDC consists of transitions from one regulatory state to another (Lewin, 1990; Murray and Kirschner, 1989). These transitions initiate the modification of substrates that determine the physical state of the cell and are themselves feedback controlled. The basic mechanisms, as elucidated by a combination of

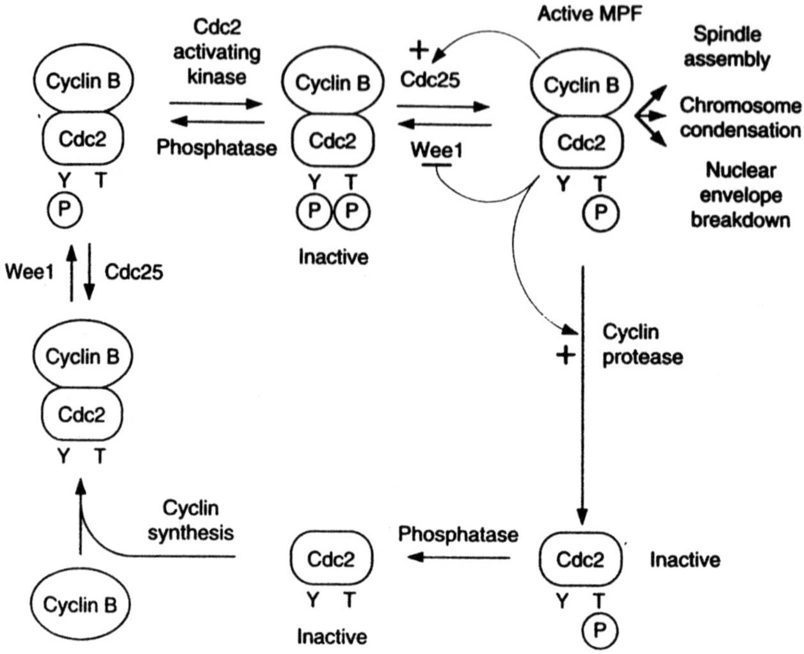

Fig. 5.1. Biochemical model of the embryonic cell cycle 'engine.' The different Cdc2/28 – cyclin B complexes formed during the embryonic cell cycle and the enzymes that act on them are shown, as well as the phosphorylation of Cdc2/28 on the tyrosine 15 (left) and threonine 161 (right) residues. At the beginning of interphase, Cdc2/28 is unphosphorylated, but then, its association with cyclin induces phosphorylation on tyrosine 15 and threonine 161 to produce preMPF. At entry into mitosis, preMPF is converted into its active form by Cdc 25, which removes the phosphate from tyrosine 15. Active MPF induces the destruction of cyclin, resulting in the inactivation of MPF. (From Murray and Hunt 1993, by permission)

genetic and biochemical analyses, appear to have been conserved in organisms ranging from unicells such as yeasts, to Xenopus, clams, and starfish eggs, and to cultured human cells (Murray and Hunt 1993).

Two crucial transition control points exist in the G_1 phase and at the G_2 /M boundary. The M phase is characterized by the activation of a kinase (MPF, or maturation [M phase]-promoting factor, named for its ability to induce cell division when injected into Xenopus oocytes), which consists of two component subunits (Fig. 5.1). One is the $p34^{cdc2}$ protein (the product of the Schizosaccharomyces pombe gene cdc2, or its homologue, CDC28, in Saccharomyces cerevisiae), which is able to phosphorylate casein and histone H1 in vitro and has maximal kinase activity at mitosis. The other is cyclin, whose cellular concentration fluctuates during the CDC: it accumulates gradually during interphase, forms a complex with $p34^{cdc2}$, and activates the protein kinase; it is then degraded during mitosis, turning off kinase and MPF activity. The system thus behaves like an oscillator, or clock, which is reset to its interphase state during mitosis (Murray and Kirschner 1989). The first cell divisions in sea urchin eggs, in frog oocytes, and in Drosophila embryos are triggered by the accumulation of cyclins above a critical concentration. Less is known about the initiation of DNA synthesis, but a complex comprising $p34^{cdc2}$, and possibly other cyclin-like molecules, may play a role (Cross et al. 1989; Murray and Hunt 1993; for compendia of recent original research articles, see Cold Spr. Harb. Symp. Quant. Biol., vol. LVI [The Cell Cycle], Cold Spring Harbor Laboratory Press, 1991; and Ciba Foundation Symposium, vol. 170 [Regulation of the Eukaryotic Cell Cycle], Wiley and Sons, Chichester 1992).

There is also strong evidence that autonomous oscillators, independent of either $p34^{cdc2}$, cyclins, or both) are part of, or are coupled to the CDC (see Edmunds 1984, 1988). Thus, when the S. pombe cell cycle is arrested by a temperature-sensitive cdc2 mutation, the activity of nucleotide diphosphate kinase and the rate of CO_2 production continue to oscillate, with a period about that of the CDC (Creanor and Mitchison 1986; Novak and Mitchison 1986). Klevecz (1976) has shown that mammalian cell cycle lengths are not smoothly distributed, but are clustered (or quantized) at intervals of 3 to 4 h. He hypothesized a quantal subcycle (G_q), driven by an autonomous clock whose period would be approximately 4 h. The duration of G_q does not vary with temperature over a range of 6-7°C, whereas cell cycle duration varies greatly (Klevecz and King, 1982).

The view, therefore, that the duration of the CDC would depend only on the accumulation of cellular components such as cyclins is overly simplistic, particularly in the case of circadian timers, which have been shown to underlie persisting cell division rhythmicity in numerous and diverse systems (Edmunds and Laval-Martin 1984), including the unicell Euglena gracilis, our research organism.

5.1.2 Cell Division Rhythmicity in Wild-Type and Mutant Euglena

The CDCs of many algae, fungi and protozoans exhibit persisting circadian rhythms of cell division, or 'hatching' (Edmunds 1978: Edmunds and Adams 1981). Division occurs only at a certain phase of the circadian cycle – often the times ('subjective' nights) in DD or LL corresponding to the dark intervals in an environmentally synchronizing LD cycle. This 'gating' phenomenon, reflecting an interaction between a circadian oscillator and the CDC, has been intensively investigated in the eukaryote Euglena. The formal properties of circadian clocks – in particular, entrainability by diurnal and non-24-h cycle (Cook and James 1960; Edmunds 1965a, 1978: Edmunds and Funch 1969a,b: Ledoigt and Calvayrac 1979) and temperature (Terry and Edmunds 1970a,b), persistence (Edmunds 1966), initiation, phase-shiftability (Edmunds et al. 1982), phase singularity (Malinowski et al. 1985), and temperature compensation (Anderson et al. 1985) – have been found to characterize the circadian rhythm of cell division in photoautotrophically cultured wild-type Euglena gracilis Klebs (strain Z) (review: Edmunds and Laval-Martin 1984).

In order to eliminate a cellular compartment in which the clock might lie and to obviate possible signaling influences of the LD cycles previously used to elicit rhythmicity and provide energy for growth, we have chosen the achlorophyllous ZC mutant of Euglena – an obligate organotroph – for a series of studies parallel to those carried out in the wild-type strain (Carré et al. 1989a). Division rhythmicity could be entrained by LD:12,12 or even by a one-pulse skeleton photoperiod (LD:1,23), and the rhythm free-ran in ensuing DD for at least 8 days with a period (τ) of about 25.5 h, providing that the overall growth rate (g) was >24 h (see Fig. 5.3). Similar, though less extensive results had been obtained earlier with other photosynthesis mutants of Euglena (Edmunds 1975, 1978; Edmunds et al. 1976; Jarrett and Edmunds 1970).

A 'strong' (Type 0) phase-response curve (PRC) for light signals in the ZC mutant strain was generated from 15 phase-shifting experiments (Carré et al. 1989a). Phase delays ($-\Delta\phi$) were obtained between Circadian Times (CT) 08-18, and phase advances ($+\Delta\phi$) between CT 18-0. The PRC for the mutant resembled that previously obtained with the wild-type strain (Edmunds et al. 1982). These studies, therefore, provide even more conclusive evidence for the role of a circadian oscillator in the control of the CDC and have effectively circumvented the problem of the dual use of imposed LD cycles: as an energy source, or 'substrate', for growth, and as a timing cue (Zeitgeber) for the underlying clock. They further demonstrate that the presence of a functional chloroplast compartment is unnecessary for circadian timekeeping in Euglena and suggest that phase-shifting the clock by light occurs via similar pathway in photosynthetic and nonphotosynthetic cell types.

Thus, the circadian oscillator(s) plays a key role in the control of the CDC in Euglena, taken to be representative of other eukaryotic microorganisms (and, perhaps, of multicellular systems). Mitosis would not be an essential part of the oscillator but would lie downstream from it: blockage of cell division should not stop the system from oscillating (Edmunds and Adams 1981; Edmunds and Laval-Martin 1984). Therefore, cell division would be a 'hand' of the underlying clock. We have tested this hypothesis in the wild type in two ways: (i) If the division rhythm (free-running in LD: 3,3) was stopped due to low initial levels of vitamin B_{12} , and if this inhibition subsequently was released by readdition of B_{12} , the rhythm started up again in phase with an unperturbed control; and (ii) If a pulse of lactate was given to a free-running culture, temporarily accelerating the CDC and overriding circadian oscillator controls (Jarrett and Edmunds 1970), the phase of the rhythm when it was finally restored after the substrate had been depleted was in phase with that of an unperturbed control. These results are consistent with those found previously for the in-phase restoration of rhythmicity in the P_4 ZUL mutant free-running in LL by the addition of sulfur-containing compounds to the medium (Edmunds et al. 1976).

5.2 Coupling between Oscillator and Cell Cycle: Role of Cyclic AMP

The mechanism whereby the circadian oscillator (and other higher frequency, ultradian oscillators) interact with the cyclin- $p34^{cdc2}$ pathway for the control of cell division is an important question that we now address. Cyclic AMP, which is known to play a pivotal role in cellular regulation (Whitfield et al. 1987), if not an element of the clock itself, may effect such a coupling between oscillator and CDC, participating (in concert, perhaps, with other players) in the gating of CDC events to specific phases of the circadian cycle.

Indeed, cAMP seems to have the capacity to control some of the rate-limiting steps of CDC progression, stimulating the proliferation of some cells, but having the opposite effect, or no effect at all on others (Dumont et al. 1989; Whitfield 1990). Such contradictory findings might be explained by a variation in different systems of the cAMP concentration required for optimal stimulation. Interestingly, transient increases of cAMP are correlated with cell cycle transitions at both the G_1 / S and G_2 / M boundaries in many cell types. Genetic experiments in Saccharomyces (Matsumoto et al. 1983), and pharmacological studies in mammalian cells (for reviews, see Boynton and Whitfield 1983; Whitfield et al. 1987) have shown that a transient rise and ensuing fall of the cAMP level are necessary for the initiation of DNA synthesis, and that a second cAMP surge is correlated with the onset of mitosis. Similar changes of cAMP concentration at different phases of the CDC have been found in Euglena (Carell and Dearfield 1982).

These signals coincide with the action(s), respectively, of $p34^{cdc2}$-p60 (G_1 cyclin) protein kinase complexes and of $p34^{cdc2}$ - cyclin B (mitosis cyclin) protein kinase complexes. Quite recently, Baroni et al. (1994) and Tokiwa et al. (1994) have demonstrated the repression of G_1 cyclin expression (through inhibition of transcription of the cyclin genes CLN1 and CLN2) by the Ras/cAMP pathway in Saccharomyces. Likewise, mitotic CDC2 kinase is negatively regulated by cAMP-dependent protein kinase in mouse fibroblast cell free extracts (Hohmann et al. 1993).

In recent studies (Carré et al. 1989b), we have shown that cultures of the achlorophyllous ZC mutant of Euglena, displaying free-running, circadian rhythms of cell division in DD (see Fig. 5.3), also exhibit bimodal circadian variations in their level of cAMP (Fig. 5.2). These oscillations in cAMP concentration could be phase-shifted by light signals in a manner predicted by the PRC previously derived for the phase-resetting of the cell division by light pulses (Carré et al. 1989a). These results strongly suggested that cAMP levels are controlled by the same endogenous clock as that regulating the CDC. Maximum cAMP levels occurred at the beginning of the light period (CT 0-2, when cells are in G_1) and at the onset of darkness (CT 12-14, corresponding to the onset of mitosis).

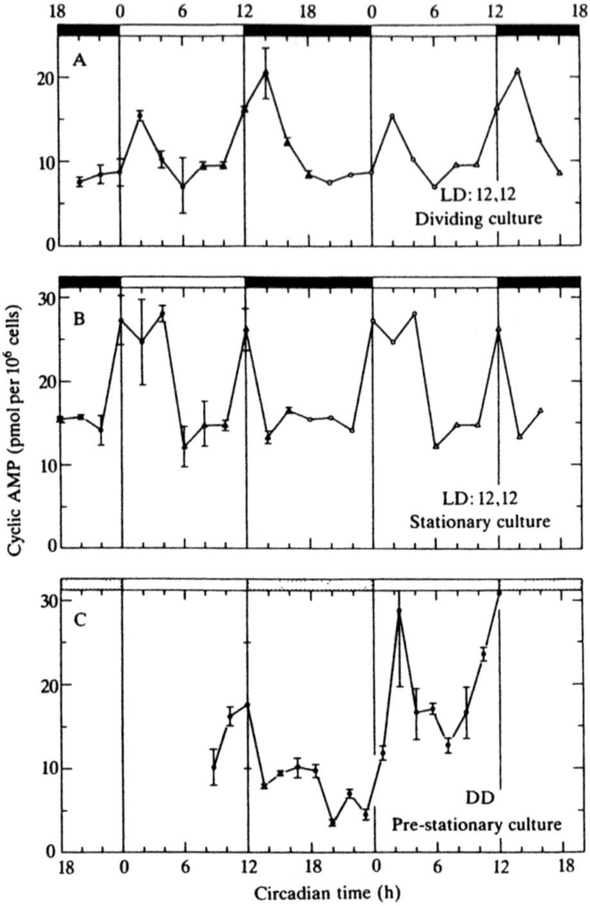

Fig. 5.2A–C. Entrained and free-running circadian oscillations of the cAMP level in synchronously dividing and in nondividing cultures of the ZC mutant of Euglena gracilis (strain Z). **A** Diurnal variations of the cAMP level in cultures exhibiting rhythmic cell division, entrained to a 24-h period by LD:12,12. **B** Persistence of the bimodal 24-h oscillation of cAMP level after the cells had ceased dividing. **C** Variations of cAMP level in a culture displaying a free-running rhythm of division in DD (τ = 27 h) and just entering the stationary growth phase. For **A** and **B**, cell extracts were prepared every 2 h over a 12-h time span from two cultures entrained by out-of-phase LD cycles, thus scanning the entire circadian period. Results were double-plotted (open symbols) as a function of the circadian time. In **C**, cell extracts were prepared every 2 h over a 34-h time span. In order to facilitate comparison with curves **A** and **B**, results were then normalized to a 24-h period. (From Carré et al. 1989b)

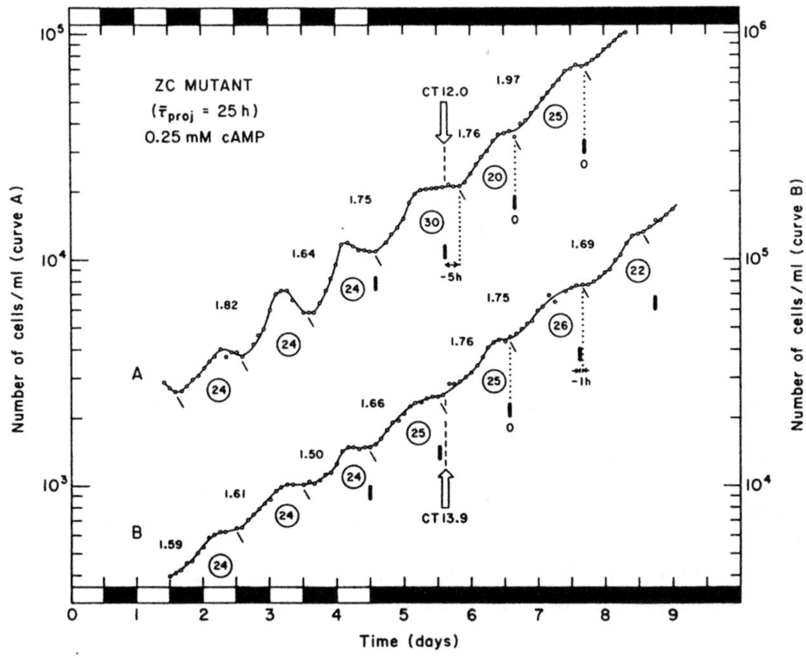

Fig. 5.3. Perturbation of the circadian rhythm of cell division by cAMP in the ZC mutant of Euglena. Cultures that had been synchronized by LD:12,12 were transferred to DD. Cyclic AMP was injected into the culture during the second circadian cycle in DD at the circadian time (CT) indicated by the open arrow. The onset of cell division was used as a phase marker (vertical dotted line) and compared to the theoretical phase of the rhythm (black markers) in an unperturbed control culture, projected from the last division step before the perturbation with a period of 25 h, a value which corresponded to the free-running period of the rhythm in DD. Phase differences are given here in real time. The period length (interval between successive division bursts) is encircled. Step size (factorial increase in cell concentration, from plateau to plateau) is given on the left of the corresponding step. (From Carré and Edmunds 1993)

At the time, these variations appeared to be CDC-independent, since they persisted after the cultures had reached the stationary phase of growth. More recent studies indicate, however, that in cells maintained for some days in stationary-phase conditions in the virtual absence of cell division, the oscillations in cAMP level continue only with greatly reduced amplitude or are absent. We believe that such a periodic, bimodal cAMP signal (initiated by the circadian oscillator, or in conjunction with it) may participate in the gating of CDC events to specific phases of the circadian cycle.

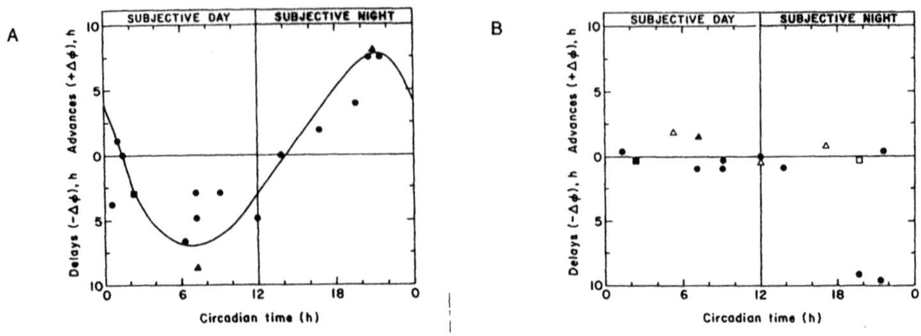

Fig. 5.4A,B. Phase-response curves for the perturbation of the free-running rhythm of cell division by cAMP in the ZC mutant of Euglena. The curves were derived from 17 experiments similar to those shown in Fig. 5. **A** Advances or delays of the first division step following cAMP injection into the culture medium. **B** Steady-state phase shifts, measured (when possible) 3 to 4 days after the perturbation. Open symbols: cell suspension was diluted 50-fold with fresh medium after a 1-h exposure to cAMP. Filled symbols: culture was not diluted. Triangles, 500 μM cAMP; circles, 250 μM cAMP; squares, 100 μM cAMP. (From Carré and Edmunds 1993)

5.2.1 Perturbation of the Cell Division Rhythm by Cyclic AMP

Thus, cAMP may play a central role as a coupling link for control of the CDC by the circadian clock, regulating the transition through the G_1 / S or the G_2 / M boundaries, or both. If this hypothesis holds true, conditions that would disturb or override the control of cAMP level by the oscillator should also cause perturbations of the cell division rhythm.

We first tested for the possible role of periodic cAMP signals in the generation of cell division rhythmicity in Euglena by examining the effects of cAMP signals on the division rhythm (Carré and Edmunds 1993). Perturbations of the cAMP oscillation by exogenous cAMP (Fig. 5.3) resulted in the temporary uncoupling of the CDC from the circadian timer, causing either a shortening or a lengthening of the CDC, depending on the phase of the CDC when the drug was applied (Fig. 5.4A). Delays of the next synchronous division step (up to 9 h) were obtained when cAMP was given between CT 03 and CT 09, and advances were obtained when cAMP was given between CT 16 and CT 22. Maximum effects were obtained at CT 06–08 and at CT 18–20, corresponding to times when endogenous cAMP levels were minimal (see Fig. 5.2). Dose-response curves were also derived at CTs when maximum + Δφ 's or - Δφ 's of the CDC had been obtained. We found that as little as 1 to 5 nM cAMP was enough to perturb cell cycle transit. The circadian oscillator, in contrast to the CDC cyclin clock, and unlike excitable tissue in which cAMP does phase-shift the output rhythm (Eskin

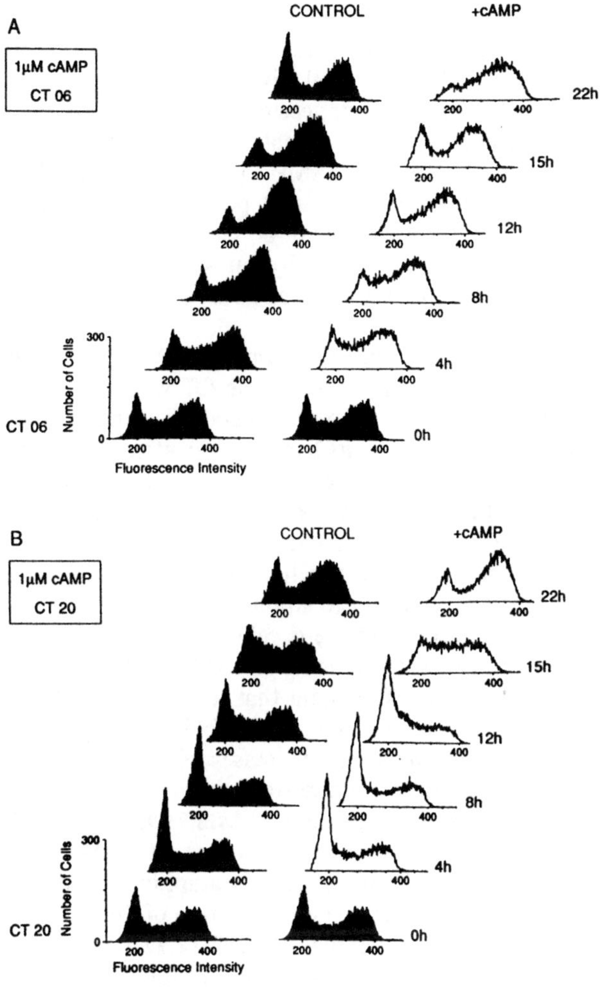

Fig. 5.5A,B. Effects of cAMP on cell cycle progression. Cultures of the ZC mutant synchronously dividing in DD were divided into two halves, one of which received cAMP (white histograms), while the other was used as a control (black histograms). Cells were harvested every 4 h following the perturbation and were fixed in 70% ethanol, treated with RNAse A, and stained with propidium iodide immediately prior to analysis in a fluorescence-activated cell sorter (FACS). **A** Addition of cAMP at CT 06 delayed the progression of cells through S phase. **B** Addition of cAMP at CT 20 stimulated mitosis (notice the decrease of the G_2 peak at t=12 h). Time elapsed since the addition of cAMP to the medium is indicated on the right of the corresponding histograms, and circadian time is given on the left. (From Carré and Edmunds 1993)

et al. 1982; Prosser and Gillette 1989), was not reset by addition of exogenous cAMP: the division rhythm returned to its original phase (Fig. 5.4B).

The advances and delays of the cell division steps observed on the growth curves following cAMP injection reflected real changes in the rate of CDC progression. Measurement of the DNA content of cells by flow cytometry (Carré and Edmunds 1993) indicated that cAMP injected between CT 06 and CT 08 delayed progression through S phase, and perhaps also through mitosis (Fig. 5.5A). When added at CT 18–20, cAMP accelerated the G_2 /M transition (Fig. 5.5B). These effects on the cell division rhythm were only transitory, and after 48 h cell division occurred at the same phase as in unperturbed controls (see Fig. 5.4B).

Thus, we have demonstrated that exogenous cAMP signals cause lengthenings or accelerations of the CDC, depending on the CT when the drug is added. We still needed to determine whether the endogenous circadian variations of cAMP level were of sufficient amplitude to have similar effects on CDC progression. Our approach was to test the effect of drugs that reduced the amplitude of the cAMP oscillation, or that kept cAMP at a level such that all cAMP receptors should have been saturated in a permanent manner. We anticipated that such drugs would prevent the expression of cell division rhythmicity.

Forskolin, which we have shown to stimulate adenylate cyclase in Euglena at times when the activity of the enzyme is minimal (Tong et al. 1991), would be expected to permanently elevate the level of cAMP. Indeed, when forskolin was injected into a culture of the ZC mutant that was maintained in LD:12,12, the cAMP level was increased by as much as threefold, as compared to an untreated control (Tong et al. 1991). Maximum increases of cAMP level corresponded to times when cAMP in the control culture was minimum; forskolin had little effect at times when cAMP was at its maximum. As a result, the cAMP level in the presence of forskolin never exceeded the range of cAMP concentrations in control cultures, but oscillated with a reduced amplitude, and at a higher level. When forskolin was added to a free-running culture of the ZC mutant (Carré and Edmunds 1993), previously synchronized by LD:12,12, at the time of its transfer to DD (at approximately CT 12), we observed a rapid desynchronization of the cell population (Fig. 5.6a). The cells appeared to go through one synchronous division step, then to resume nearly exponential growth. Very low-amplitude waves of cell division were still visible, however, for a minimum of three more days, a sign that that the circadian timer was still running unaffected. Such a rapid loss of cell division rhythmicity sometimes occurs after transfer of control cultures to DD, but heretofore has never been observed in cultures that had been maintained in 24-h LD cycles. When forskolin was added to a culture that was entrained by LD:12,12, a rapid damping of the cell division rhythm also was observed (Fig. 5.6b). This effect was not phase-dependent: similar results were obtained when the drug was added at CT 12 (Fig. 5.6a,b) or at CT 04, 08, 15, or 00 (not shown). These results suggest that the amplitude of the cAMP oscillations determine the amplitude of the cell division rhythm and are consistent with the hypothesis that cAMP acts downstream from the oscillator – perhaps in a parallel pathway, as Zatz (1992) has suggested for chick pineal cells – and that the cAMP

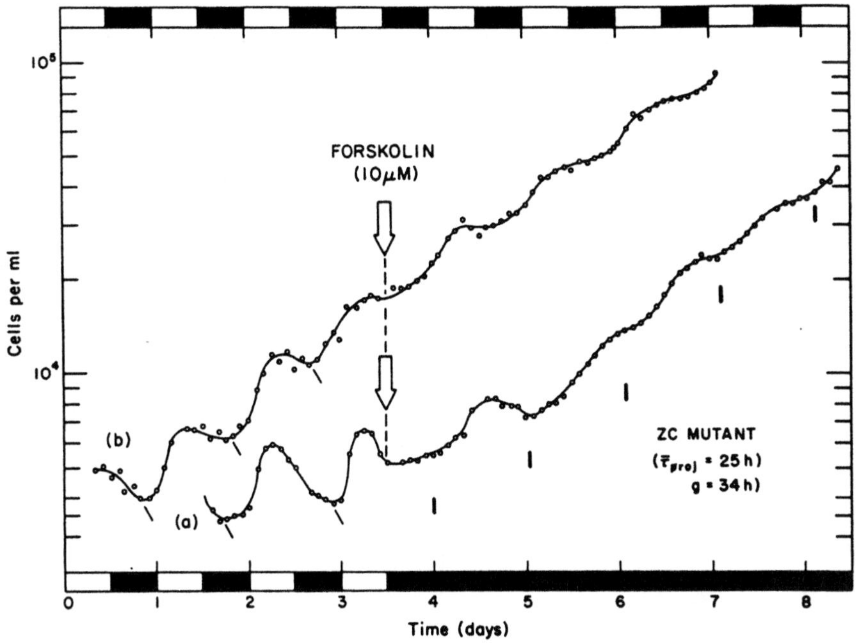

Fig. 5.6. Effect of forskolin on the cell division rhythm in the ZC mutant. Cell division was synchronized by LD:12,12 prior to the addition of forskolin. a, culture transferred to DD at the time of the addition of the drug (indicated by the open arrow); b, culture maintained in LD:12,12. Both experiments show a rapid loss of cell division rhythmicity, but the generation time was unaffected. (From Carré and Edmunds 1993)

oscillation is an essential component of the signaling pathway for the control of the CDC by the circadian oscillator.

5.2.2 Upstream Analysis: Genesis of the Oscillation in Cyclic AMP

What upstream controls regulate the level of cAMP? Key factors in the cAMP metabolic pathway are two enzymes responsible for its generation and degradation: adenylate cyclase (AC) and phosphodiesterase (PDE). In LD:12,12 these enzymes were found to undergo bimodal, circadian variation of activity in both dividing and nondividing cultures of the ZC mutant (Fig. 5.7), although more recent results suggest that the amplitudes of the rhythms are attenuated in later stationary-phase populations (Tong et al. 1991). Maximal AC activity occurred 2h after the onset of light (CT 02) and at the beginning of darkness (CT 12-14); these times corresponded to the acrophases for the rhythmic changes in cAMP content that have been previously reported (Carré et al. 1989b). The activity of PDE also exhibited a daily oscillation, but with an inverse phase

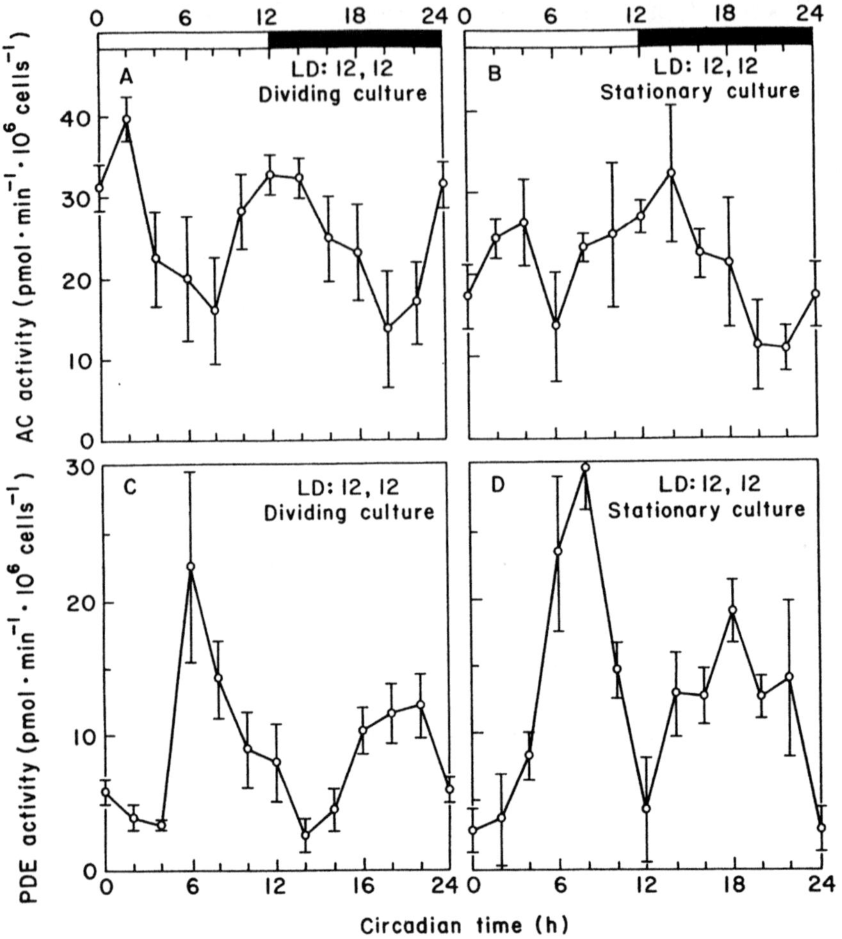

Fig. 5.7A–D. Bimodal circadian rhythms of **A,B** adenylate cyclase (AC) activity and **C,D** phosphodiesterase (PDE) activity in either **A,C** synchronously dividing or **B,D** stationary-phase cultures of the ZC mutant of Euglena gracilis. For experiments **A** and **C** the results shown are averages of triplicate assays from two different cultures, and for **B** and **D** the data are averages of triplicate assays from a single culture. (From Tong et al. 1991)

pattern. Both the AC and PDE activity rhythms persisted in DD. The activity of AC was activated significantly in vivo by forskolin (a highly specific activator of AC that can fully stimulate AC activity directly, independently of G proteins) at the trough phase (CT 20), as found also in rat heart (Lemmer et al. 1986), while that of PDE was inhibited by 3-isobutyl-1-methylxanthine (IBMX) at its peak phase. These results indicate that the rhythms of AC and PDE are key factors in generating the circadian oscillations of cAMP content in Euglena, which appear to be under control of an endogenous pacemaker.

The activity of AC was activated significantly in vivo by forskolin (Tong et al. 1991), a highly specific activator of AC that can fully stimulate AC activity directly, independently of G proteins. Forskolin stimulation of AC did not result in uniform effects throughout the 24-h day but varied considerably with circadian time. The degree of AC potentiation by an in vivo forskolin pulse appeared to be inversely dependent on the basal level of AC: at its trough phase (CT 20), AC activity was significantly stimulated by as much as 114% (bringing it to exactly the same level as that found at the peak phase), while at the peak phase (CT 12), only a 10% stimulation was observed. The same result has been demonstrated by Lemmer et al. (1986) in rat heart: The level of cAMP was significantly increased by the addition of forskolin (5 mg/kg) at CT 12 in LD:12,12, but the effect disappeared at CT 20 under the same experimental conditions. Thus, since forskolin appears to stimulate AC activity at different circadian times to the same maximal value regardless of its initial basal level, we conclude that the oscillatory activity of the enzyme derives from modulatory cellular effectors emanating from the circadian pacemaker rather than by changes in the amount of the enzyme protein itself.

While previous studies have emphasized the role of AC, along with cAMP, in signal transduction and neoplastic transformation (Whitfield 1990), less attention has been paid to the role of PDE. PDE is an important component of the signal system because it destroys cAMP after a pulse of synthesis and, thereby, permits the cell to recover from a refractory state induced by cAMP. We found that addition of 50 μM IBMX to intact cells at CT 20, corresponding to the peak of the rhythm in PDE activity (see Fig. 5.7), depressed PDE activity by about 51%, while at CT 12 (the trough phase) IBMX had only a slight (8%) effect (Tong et al. 1991). The hydrolysis of cAMP by cell extracts is catalyzed by at least five types of PDE isoenzymes: calmodulin-stimulated PDE, cyclic GMP (cGMP)-stimulated PDE, cGMP-specific PDE, low K_m PDE, and nonspecific PDE (Beavo 1988). IBMX is a nonspecific PDE inhibitor, which can affect the activity of calmodulin-stimulated PDE and cGMP-inhibited PDE 20-25 times as much as that of cGMP-stimulated PDE in mammalian cells (Beavo 1988). Thus, the inhibitory effects of IBMX on PDE observed in Euglena extracts may be the resultant of its action on several PDE species. Furthermore, the differing effects of IBMX on PDE activity at CT 12 and CT 20 showed that the degree of inhibition was circadian-time-dependent and that IBMX had little or no effect on PDE at CT 12 (corresponding to the time when the activity was at a minimum). At CT 20 (when PDE activity was at its peak), IBMX diminished the activity to the same level as that observed at CT 12. This suggests that the relative amounts of the different types of PDE vary with circadian time and that a specific PDE (or a specific subset of enzymes that are inhibited by IBMX) is responsible for the oscillation.

In sum, these results indicate that the rhythms of both AC and PDE may be the main factors in generating the circadian oscillations of cAMP content in Euglena, which appear to be under control of an endogenous pacemaker. We are now extending this upstream analysis towards the clock, examining the possible roles

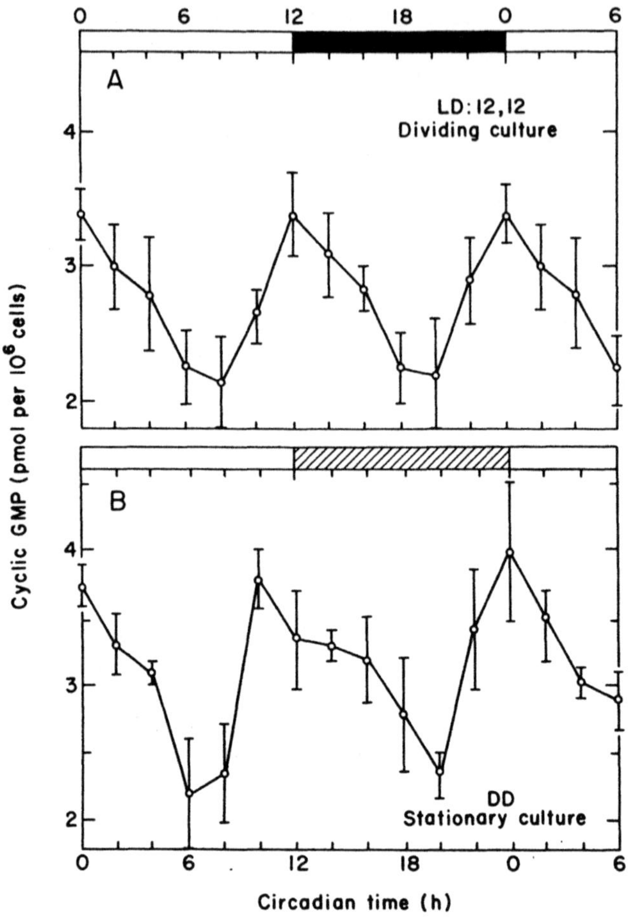

Fig. 5.8A,B. Circadian rhythms of cGMP in **A** dividing and **B** nondividing cultures of the achlorophyllous ZC mutant of Euglena gracilis Klebs (strain Z), grown in LD:12,12 and DD, respectively. The results shown are averages of two independent experiments with duplicate assays; error bars indicate the range of the values. The hatched bar in **B** indicates subjective night in order to facilitate comparision with **A**. (From Tong and Edmunds 1993)

of Ca^{2+}, calmodulin, inositol 1,4,5-triphosphate (IP_3), and cyclic GMP (cGMP) in regulating the activities of AC and PDE (Edmunds and Tamponnet 1990). The 'Yin Yang,' or dualism hypothesis proposed about 20 years ago (Goldberg and Haddox 1977) provided a theory of biological regulation through the opposing actions of cAMP and another second messenger, cGMP. According to this theory, reciprocal changes in the levels of the two cyclic nucleotides should bring about a maximum expression of the mitogenic signal. An interesting observation was that an increase in tissue cGMP levels was accompanied by either a decrease, increase, or no change in cAMP concentration. The question then arises as to the origin of the cAMP variation and if these changes in cGMP/cAMP ratios are stochastic, or whether, as the hypothesis posits, cGMP may serve as a positive mediator.

To determine if cGMP plays a role in the mediation of circadian rhythmicity of the AC-cAMP-PDE system in the ZC mutant of Euglena, the levels of cAMP and

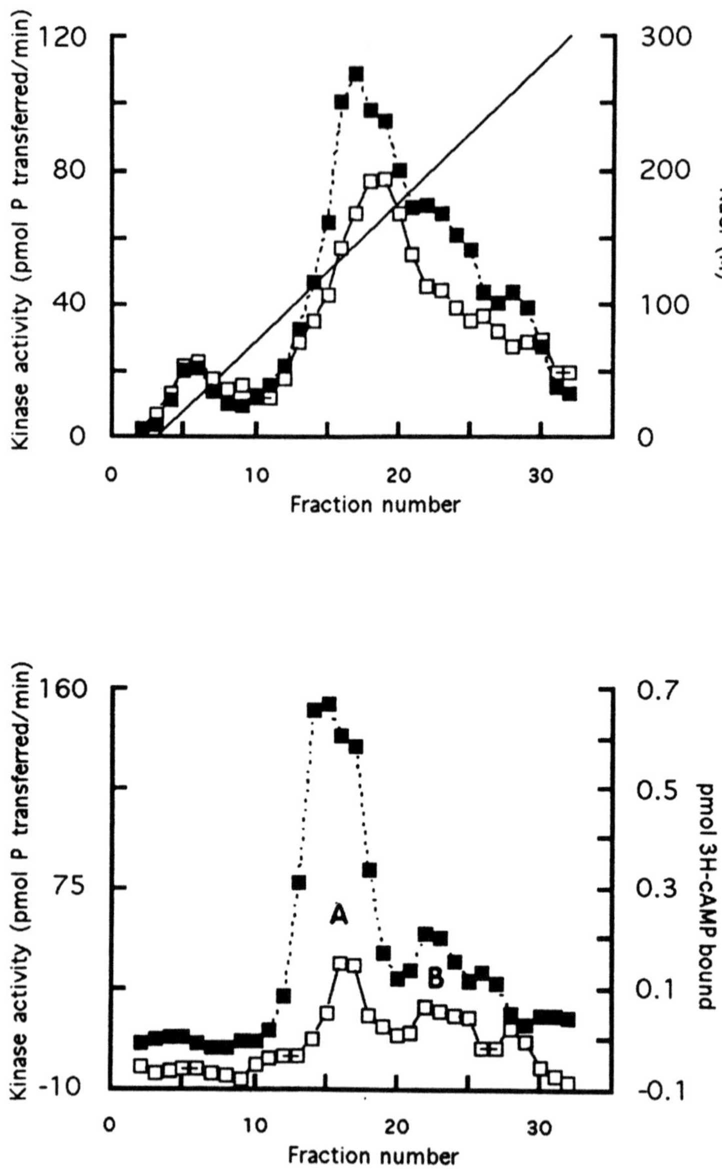

Fig. 5.9. Separation of cAMP-dependent kinases and of cAMP-binding proteins by DEAE-cellulose chromatography. Top: the basal kinase activity (open symbols) was stimulated in the presence of cAMP (filled symbols). Bottom: after subtraction of the basal activity, the cAMP-stimulated kinase activity (open symbols) was resolved into two peaks (A and B), which overlapped with two peaks of cAMP-binding activity (filled symbols). From Carré and Edmunds 1992)

Table 5.1. Cyclic AMP-analog specificities of cAMP-dependent kinases from the ZC Mutant of Euglena and minimum doses of cAMP analogs that perturb the cell division rhythm at CT 06-08 or at CT 18-20 [a] .

Analog[b]	(1) cPKA ($\underline{K}_a$, nM)	(2) CT 06–08 (Delays)	(3) cPKB ($\underline{K}_a$, nM)	(4) CT 18–20 (Advances)
Cyclic AMP	5 (200%)	1	50 (200%)	50
8-CPT-cAMP	no effect	no effect	2 (150%)	0.1
8-BZT-cAMP	200 (600%)	100	no effect	no effect
6-MBP-cAMP	5 (130%)	10	50 (130%)	ND

[a]The magnitude of the stimulation of the kinases (% of basal activity) by the cAMP analogs is given in parentheses (columns 1 and 3). The maximum concentration tested was 1 mM. The lowest concentrations of the cAMP analogs that were found to reproduce the effects of 250 nM cAMP applied at either CT 06–08 (column 2) or at CT 18–20 (column 4) are given. The maximum concentration tested was 1 µM. ND, not determined. (From Carré and Edmunds 1993).
[b]8-CPT-cAMP, 8-(4-chlorophenylthio)-cAMP; 8-BZT-cAMP, 8-benzylamino-cAMP; 6-MBP-cAMP, $\underline{N}^6$-Monobutyryl-cAMP.

To our knowledge, this is the first demonstration of a cAMP-dependent kinase in a photosynthetic organism. Thus, cAMP signalling in plant cells may proceed by the phosphorylation of target proteins by cAMP-dependent kinases, in a manner similar to that of animal cells.

A correlation between a cAMP-analog's potency in activating one type of kinase and in causing physiological responses can provide evidence for a role of this enzyme in mediating the effect of cAMP. The differential activation of the two kinases identified in Euglena extracts by these cAMP analogs provided us with a tool for the study of their respective roles in the control of cell cycle progression: we determined the mimimum doses of cAMP or of cAMP analogs that caused perturbations of the cell division rhythm during the subjective day, or during the subjective night (Carré and Edmunds 1993). Interestingly, different results were obtained at CT 06-08, and at CT 18-20, suggesting that the effect of cAMP at the different CTs was mediated by two different cAMP kinases (Table 5.1). Thus, 8-benzylamino-cAMP (8-BZA-cAMP), which selectively activates cPKA, induced delays of cell division when added at CT 06-08, but had no effect at CT 18–20. Reciprocally, 8-(4-chlorophenylthio)-cAMP (8-CPT-cAMP), which specifically activates cPKB, induced early cell division and a loss of division rhythmicity when added at CT 18-20 but did not perturb the cell division rhythm when added at CT 06-8. Moreover, there was a correlation between the doses of cAMP, 8-BZA-cAMP, 8-CPT cAMP, and $\underline{N}^6$-monobutyryl-cAMP (6-MBT-cAMP) that caused perturbations of the CDC at CT 06-8 and the $\underline{K}_a$ values of cPKA for these analogs, a result that suggests that cPKA mediates the delaying effects of cAMP at those CTs. Similarly, there was a correlation between the doses of the same nucleotides that caused perturbations of the CDC at CT 18-20 and the $\underline{K}_a$ of cPKB for these analogs, an observation indicating that cPKB mediates the accelerating effects of cAMP at CT 18-20. A simple explanation for these findings would be that these kinases are expressed at different stages of the CDC, as has been previously described for type I and type II cAMP-dependent kinases from mammalian cells.

5.2.4 Model for Circadian Control of the Cell Division Cycle

These results can be incorporated into a model (Fig. 5.10) for the coupling of the CDC to the circadian oscillator (Carré and Edmunds 1993). We propose that the cAMP surge at CT 00-02 delays DNA synthesis and holds the cells at a restriction point in G_2 to prevent cell division during the subjective day; the cells are released from this blockage after cAMP levels subside, and the G_2 / M transition, or mitosis itself, is accelerated by the second cAMP peak, at CT 12-14, so that mitosis is phased to the subjective night. The delaying effects of cAMP on cell cycle progression during the subjective day would be mediated by the activation of cPKA, and the stimulation of mitosis during the subjective night by the activation of cPKB. Activation of either of these kinases would cause the phosphorylation of a different set of targets and perturb different cell-cycle control pathways. cPKA and cPKB may be expressed at different phases of the CDC. Alternatively, the level of these enzymes might exhibit circadian variations, with cPKA's being expressed during the subjective day and cPKB during the subjective night. Another possibility is that the level of one of their downstream targets oscillates, so that only cPKA activation has an effect on cell cycle progression during the subjective day, and cPKB during the subjective night.

How does the circadian oscillator, acting through (or in conjunction with) the AC-cAMP-PDE system and the cAMP-dependent kinases (cPKA and cPKB) that we have identified, ultimately couple to the MPF clock regulating the CDC? Preliminary western blot analyses of protein extracts from dividing and stationary cultures of the ZC mutant in DD (Fig. 5.11B) confirm the presence of a homolog of $p34^{\underline{cdc2}}$, the catalytic subunit of MPF, reported earlier in wild type Euglena and in other algae (John et al. 1989, 1993) and suggest that its abundance throughout the CDC is invariant, decreasing to a much reduced level when cells cease dividing (Mohabir and Edmunds 1994, unpublished results). Further, a relative shift in electrophoretic mobility of the monomer was observed to occur in a cycle-dependent manner: a more slowly moving band appeared at CT 04 and disappeared at CT 12. Such changes in mobility have been attributed to phosphorylation and dephosphorylation. These initial findings, if replicated, suggest that the activity of this polypeptide depends on posttranslational mechanisms, and, thus we have turned our attention to the regulatory subunit(s) of MPF. On the assumption that cyclin B is a likely candidate, we have begun western blot analyses using monoclonal antibodies (Fig. 5.11A) and have found that, in contrast to $p34^{\underline{cdc2}}$, the abundance of cyclin B in protein extracts varied quantitatively during both the circadian cycle and the CDC, being detected only at CT 20 and CT 24 (00). Further characterization of this key player in CDC regulation would relate to its state of phosphorylation, which can be determined

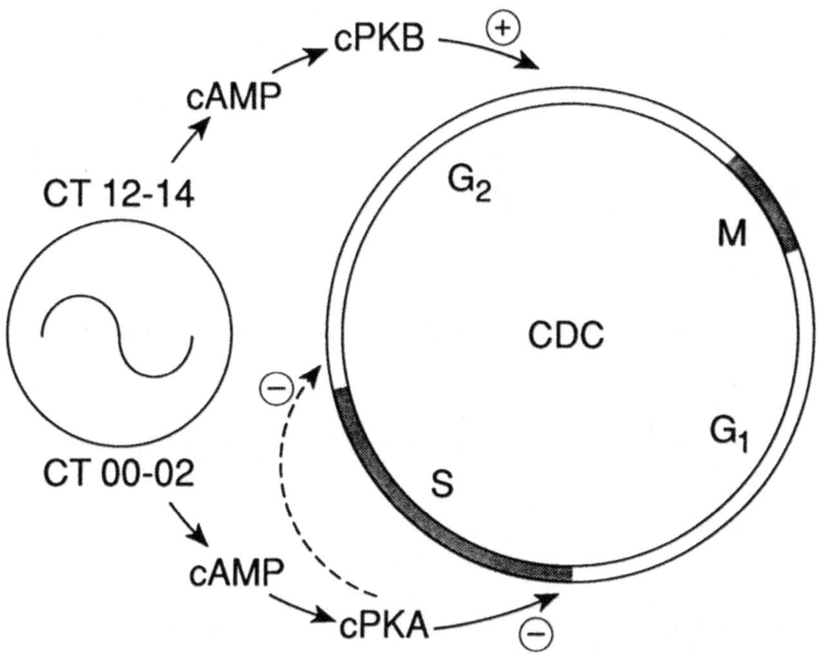

Fig. 5.10. Model for the 'gating' of cell division by the circadian oscillator. We propose that the cAMP surge at CT 00-02 delays DNA synthesis (and, perhaps, holds the cells at a restriction point in G_2) to prevent cell division during the subjective day. The cells are released from this (these) blockage(s) after cAMP levels subside, and mitosis is stimulated by the second cAMP peak (at CT 12-14) so that cell division is phased to the subjective night. Opposite effects of cAMP on cell cycle progression are explained by its action through two different cAMP-dependent kinases (cPKA and cPKB), which would be expressed at different stages of the CDC (or at different phases of the circadian cycle) and which have different sets of targets. (From Carré and Edmunds 1993)

by phosphorylation in vivo with ^{32}P followed by immunoprecipitation, western blot analysis and autoradiography. Comparison of extracts from dividing and nondividing cells at different CTs in DD should help dissect CDC-dependent and circadian clock-driven components.

5.3 Problems and Prospects: Interface with the Cyclin Clock

How does cross-talk take place between the circadian oscillator and the cyclin clock? In the preceding sections, we have traced the flow of information along the signal transduction pathway via the AC-cAMP-PDE system and the cAMP-

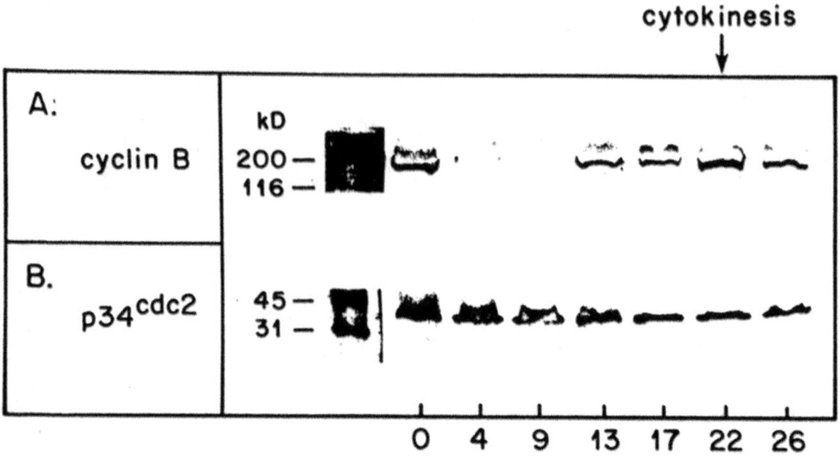

Fig. 5.11A,B. Relative abundance of homologues of maturation-promoting factor (MPF) at different circadian times in $(NH_4)_2\ SO_4$ fractions of synchronously dividing cultures of the ZC mutant of Euglena. Cyclin B, the regulatory subunit of the 34 kD protein, varied quantitatively during both the circadian cycle and the CDC (**A**), being detected only at CT 20 and CT 24, whereas the abundance of the catalytic subunit ($p34^{cdc2}$) was cycle-invariant, although a relative shift in electrophoretic mobility of the monomer did occur in a cycle-dependent manner (**B**), perhaps reflecting (de)phosphorylation. (Mohabir and Edmunds 1994, unpublished results)

dependent kinases (cPKA and cPKB) that we have identified, which we believe ultimately couple to the MPF clock regulating the CDC (see Fig. 5.10).

Mitotic kinase activity is dependent on the association of the two subunits of MPF and the phosphorylation state of the complex. Association of cyclin B (the regulatory subunit) with the $p34^{cdc2}$ (catalytic) subunit is required for mitotic kinase activity, but only after dephosphorylation of the latter on tyrosine residue 15 by a phosphatase (cdc25 in fission yeast) is that activity expressed (Jessus and Beach 1992). Kinase activity promotes the degradation of cyclin B and thus is itself nullified, thereby prempting catastrophic mitotic arrest (Félix et al. 1990). Rephosphorylation of tyrosine 15 occurs after mitosis by a tyrosine kinase (TK), the wee1 gene product of fission yeast. In initial studies, we have confirmed the presence of the homologues of $p34^{cdc2}$ and cyclin B (as well as wee1 product) in the achlorophyllous ZC mutant of Euglena and have demonstrated that the level of the former is CDC-invariant while that of the latter rhythmically varies across the CDC and the circadian cycle (see Fig. 5.11). We plan to assay for histone H1 kinase activity of MPF after purification on p13-agarose conjugate.

We also have begun a further characterization of the nature of the changes in electrophoretic mobility that we have observed (see Fig. 5.11B). Remarkably, these cyclical gel-shifts in Western blots of cdc2 from nondividing cells appear to persist in DD, occurring at circadian times corresponding to the G_2/M transitions had the CDC continued under these free-running conditions. Pre-treatment of samples with potato acid phosphatase obliterates the electrophoretically less mobile band and the antiphosphotyrosine immunoreactivity of cdc2. Further, sequential blotting with phosphotyrosine antibody followed by anti-PSTAIR has shown that the dephosphorylation was that of tyrosine residues of cdc2. These preliminary findings justify our earlier conceptualization that for circadian precision the clock should couple with its overt output rhythm at a point at or near the final execution step – G_2 / M, in the case of the CDC. Our results also underscore the importance of the kinase and phosphatase that phosphorylate and dephosphorylate, respectively, the tyrosine residue of cdc2. We have immuno-identified a weel homolog that shows an invariant CDC and circadian profile of its level in Euglena. It is imperative that a homolog of cdc25 also exist in this eukaryote, and we will screen for its presence. If we are successful in identifying this key phosphatase 'player,' we would then determine its level, activity and state of phosphorylation during the CDC in dividing cultures and, subsequently, across the circadian cycle in stationary-phase populations held in DD to ascertain if it is CDC-independent and clock-controlled. Complete amino acid sequences of cdc25 homlogs in frog, yeast, man and fruit fly have been published, but a highly conserved domain such as the PSTAIR region of cdc2 has yet to be identified and used as an antigenic peptide for commercial production of an antibody capable of reliable detection of cdc25 in different species. Thus, we may have to consider raising an antibody to the whole protein from Euglena itself. Purification of the protein for this purpose may be simplified, since theoretically it should be possible to co-immunoprecipitate it with antibody to either subunit of MPF, or to co-purify with p13-agarose conjugate at CDC times when the proteins are physically associated.

Western analyses have revealed a homolog of cyclin B1 in Euglena by the use of a monoclonal antibody against the human protein. The level of the protein varies in dividing cells in a CDC-dependent manner but ceases to do so in cells that have reached the stationary phase. Rather, the clb1 protein remains elevated throughout the nondividing phase (Mohabir and Edmunds, unpublished). We believe that the degradative pathway for cyclin depends on the presence of active MPF, as previously demonstrated in Xenopus interphase cell-free extracts (Félix et al. 1990), and perhaps also on the bimodal fluctuation of cAMP above certain threshold levels, which ceases to prevail as the stationary phase progresses with a concommitant decline of cAMP (see below). We plan to investigate this phenomenon and the role of cAMP in the regulation of the activities of CDC proteins at the biochemical level by using cell-free extracts taken at different phases of the CDC and circadian cycle. This will allow us to extend the studies on the stability of cyclin B in interphase extracts (Félix et al. 1990) to that in nondividing cells. We will then attempt to correlate peak and trough levels of cAMP in Euglena in DD with the levels, activities and phosphorylation status of CDC regulatory proteins – especially the subunits of MPF – in actively dividing cells. We propose to carry out experiments with homogenates treated with cAMP, with and without inhibitor of cAMP-dependent kinase (PKI), and assayed for MPF activity (with histone H1 as substrate). Should the results so indicate, we would proceed to immunoanalyze identically treated extracts to determine further the mode of action of the cyclic nucleotides with regard to direct or indirect modification of the CDC protein we have already identified. (We routinely monitor the phosphorylation state of the catalytic subunit of MPF (cdc2) by gel shift in Western blots.) Also relevant is the observation of down regulation of cdc2 kinase activity by cAMP in mouse fibroblast cell-free extracts (Hohmann et al. 1993), which was hypothesized to occur via its effect on a small, acidic, heat-stable protein consistent with inhibitor-1 of protein phosphatase 1 (PP 1). This down regulation was associated with changes in size and/or charge density of the cdc2 kinase complex in non-denaturing PAGE. We will extend this study by using our denaturing PAGE in order to elucidate the nature of the change(s) in cdc2 and more precisely determine the nature of the putative inhibitor and its modulation by cAMP via cAMP-dependent kinases.

Thus, the signal transduction pathway between cAMP and cdc2 may be envisaged to begin with cAMP's (-dependent kinase) affecting the inhibitor of PP 1 that then controls PP 1 (Hohmann et al. 1993). Phosphorylation status determines the activity of cdc25 (Izumi et al. 1992; Zheng and Ruderman 1993), which in turn regulates the status of tyrosine phosphorylation and the activity of cdc2. It is known that cdc25 is phosphorylated by cdc2 in a positive feedback loop (Hoffman et al. 1993), but its dephosphorylation by protein phosphatases 1 and 2A has been demonstrated only in vitro (Izumi et al. 1992). Compatible with schema is the recent report that PP 1 is itself an in vivo target of MPF, inhibitorily phosphorylated on Thr-320 (Dohadwala et al. 1994). This would also help to explain the dramatic amplification of MPF activity at the onset of mitosis: in addition to the phosphorylation of cdc25 by MPF by a single auto-amplification loop, the inhibitory phosphorylation of PP 1 (the proposed phosphatase of cdc25) would allow maximal phosphorylation and activity of cdc25 (the activator of MPF). Thus, if it can be confirmed that this phosphatase indeed is responsible for in vivo dephosphorylation of cdc25, the above scheme would be supported. We will test this possibility in vitro by the use of okadaic acid, microcystin, or both, which should inhibit PP 1, resulting in hyperphosphorylation and activity of cdc25, and subsequently of cdc2. Then, we would pulse cultures prior to histone H1 kinase assay and Western analysis with antibodies to cdc2, phosphotyrosine and cdc25.

Our observation of cyclic phosphorylation changes in cdc2 that persist in nondividing cells under free-running conditions raises the possibility that reversible protein phosphorylation might mediate circadian control of the CDC. Our initial work with monoclonal phosphotyrosine antibody also has revealed that, in addition to the persistence of cyclic tyrosine phosphorylation of cdc2, there are three other prominent anti-phosphotyrosine-stained bands, one of which exhibits similar circadian rhythmicity. At this stage it is not clear as to the significance of this particular protein, but reversible protein phosphorylation recently has been implicated in the mechanism of the circadian clock itself. Thus, Edery et al. (1994) have reported persistent rhythmic phosphorylation changes in the PER protein of Drosophila under constant conditions, while Comolli et al. (1994) have shown that the inhibitor of protein phosphorylation 6-DMAP (6-dimethylaminopurine), which reversibly blocks cell division in Gonyaulax polyedra, also affects the circadian expression of bioluminescence in this marine dinoflagellate. Recently, cdk- and cyclin-like proteins have been identified in post-mitotic tissues of the isolated eye of the gastropod Bulla gouldiana that are affected by treatments that phase shift the circadian rhythm of ocular membrane potential (Krucher and Roberts 1994; Roberts et al. 1992). Further work may well reveal a close similarity, if not a shared common mechanism, between the eukaryotic CDC and the circadian clock.

We also plan to investigate the role of cAMP during the S phase, which coincides with the first cAMP peak of the bimodal circadian rhythm in Euglena. To do so we will modify our earlier flow cytometric analyses (Carré et al. 1993) to include assays on cultures pulsed with inhibitors of DNA synthesis to determine the constancy of the time interval between the onsets of S and M, as has been reported for Gonyaulax (Homma and Hastings 1989). If circadian control of S phase is implicated, we then will examine the relationship between the onset of S and the cAMP peak accompanying it by pulse-labeling cells with bromodeoxyuridine (Blomhoff et al. 1988) to be able to monitor their progress after subsequent pulses of cAMP and forskolin. Recent advances in flow cytometric analysis (Gong et al. 1993) will allow us to monitor cyclin B as well as DNA content. This will be done by fixing the cells in 1:1 (v/v) solution of acetone in absolute ethanol for 30 min at -20 °C followed by incubation in primary antibody (cyclin B) and secondary antibody (fluorescein isothiocyanate-conjugated goat anti-mouse IgG). Cyclin B-labelled cells would then be stained for DNA with propidium iodide prior to measurement.

Finally, we have observed during certain experiments that require extended time spans (as, for example, in pulse experiments that must allow for transients to subside) that the levels of cAMP decline appreciably as stationary phase is prolonged. We are attempting, therefore, to devise a system that may allow us to uncouple the cell cycle from the circadian clock while the cAMP level is still typical of that in dividing cells. Since it is known that the cAMP level reflects carbon nutrient status in yeast (Russell et al. 1993) and that yeast cells may be forced into the stationary phase by carbon as well as by nitrogen starvation, we will attempt to use nitrogen deprivation as a means of arresting cell division in Euglena (stationary cultures of which can be made to resume division by the mere addition of ethanol). Alternatively, we may rely on our earlier discovery that the supply of vitamin B_{12} can be manipulated to arrest and release the CDC of Euglena without affecting the circadian oscillator (Edmunds and Laval-Martin 1984). If these methods are successful we would be able to use them to elucidate more clearly than was possible at physiologically lowered cAMP levels the relationship among the circadian clock, cAMP, and the CDC. We are particularly interested in determining whether cAMP is on a direct pathway from the clock to the CDC or whether it lies on some sort of parallel pathway as apparently it does in the chick pineal (Zatz 1992) and whether other rhythms share the same steps. In the case of a shared pathway, the other outputs might also be suppressed when cAMP declines to the relatively low levels seen in advanced stationary phase. We would include, therefore, concurrent monitoring in Euglena of other overt circadian rhythms such as the settling and phototaxis rhythms we have observed in wild-type cultures (Edmunds 1982, 1988).

References

Alvarez R, Daniels V (1990) A single column method for the assay of adenylate cyclase. Analyt Biochem 187: 94-97

Anderson RW, Laval-Martin DL, Edmunds LN Jr (1985) Cell cycle oscillators: temperature compensation of the circadian rhythm of cell division in Euglena. Exp Cell Res 157: 144-158

Baroni MD, Monti P, Alberghina L (1994) Repression of growth-regulated G1 cyclin expression by cyclic AMP in budding yeast. Nature 371: 339-342

Beavo JA (1988) Multiple isozymes of cyclic nucleotide phosphodiesterase. In: Greengard P, Robison GA (eds) Advances in second messenger and phosphoprotein research, vol 22. Raven Press, New York, pp 1-38

Blomhoff HK, Blomhoff R, Stokke T, Davies CD, Brevik K, Smeland EB, Funderud S, Godal T2 (1988) cAMP-mediated growth inhibition of a B-lymphoid precursor cell line Reh is associated with an early transient delay in G_2 /M, followed by an accumulation of cells in G_1. J Cell Physiol 137: 583-587

Boynton AL, Whitfield JF (1983) The role of cyclic AMP in cell proliferation: a critical assessment of the evidence. Adv Cyclic Nucleotide Res 15: 193-294

Brizuela L, Draetta G, Beach D (1989) Activation of human CDC2 protein as a histone H1 kinase is associated with complex formation with the p62 subunit. Proc Natl Acad Sci USA 86: 4362-4366

Buetow DE (ed) (1968a,b; 1982) The Biology of Euglena, vols I,II, III. Academic Press, New York

Carell EF, Dearfield KL (1982) A relationship between adenosine 3'-5'-cyclic monophosphate levels and deoxyribonucleic acid synthesis in Euglena. Life Sci 31: 249-254

Carré I, Oster AS, Laval-Martin DL, Edmunds LN Jr (1989a) Entrainment and phase shifting of the circadian rhythm of cell division by light in cultures of the achlorophyllous ZC mutant of Euglena gracilis. Curr Microbiol 19: 223-229

Carré I, Laval-Martin DL, Edmunds LN Jr (1989b) Circadian changes in cyclic AMP levels in synchronously dividing and stationary-phase cultures of the achlorophyllous ZC mutant of Euglena gracilis. J Cell Sci 94: 267-272

Carré IA, Edmunds LN Jr (1992) cAMP-dependent kinases in the algal flagellate Euglena gracilis. J Biol Chem 267: 2135-2137

Carré IA, Edmunds LN Jr (1993) Oscillator control of cell division in Euglena: cyclic AMP oscillations mediate the phasing of the cell division cycle by the circadian clock. J Cell Sci 104: 1163-1173

Cobb CE, Corbin JD (1988) Purification of cAMP-free and cAMP-bound forms of bovine heart cAMP-dependent kinase holoenzyme. Methods Enzymol 159: 202-208

Comolli J, Taylor W, Hastings JW (1994) An inhibitor of protein phosphorylation stops the circadian oscillator and blocks light-induced phase shifting in Gonyaulax polyedra. J Biol Rhythms 9: 13-26

Cook JR, James TW (1960) Light-induced division synchrony in Euglena gracilis var. bacillaris. Exp Cell Res 28: 524-530

Creanor J, Mitchison JM (1986) Nucleotide diphosphokinase, an enzyme with step changes in activity during the cell cycle of the fission yeast, Schizosaccharomyces pombe. J Cell Sci 86: 207-215
Cross F, Roberts J, Weintraub H (1989) Simple and complex cell cycles. Annu Rev Cell Biol 5: 341-395
Dohadwala M, da Cruz e Silva EF, Hall FL, Williams RT, Carbonaro-Hall DA, Nairn AC, Greengard P, Berndt N (1994) Phosphorylation and inactivation of protein phosphatase 1 by cyclin-dependent kinases. Proc Natl Acad Sci USA 91: 6408-6412
Doskeland SO, Ogreid D (1988) Ammonium sulfate precipitation assay for the study of cyclic nucleotide binding to proteins. Meth Enzymol 159: 147-151
Dumont JE, Jauniaux J-C, Roger PP (1989) The cyclic AMP-mediated stimulation of cell proliferation. Trends Biochem Sci 14: 67-71
Edery I, Zwiebel LJ, Dembinska ME, Rosbash M (1994) Temporal phosphorylation of the Drosophila period protein. Proc Natl Acad Sci USA 91: 2260-2264
Edmunds LN Jr (1965a) Studies on synchronously dividing cultures of Euglena gracilis Klebs (strain Z). I. Attainment and characterization of rhythmic cell division. J Cell Comp Physiol 66: 147-158
Edmunds LN Jr (1965b) Studies on synchronously dividing cultures of Euglena gracilis Klebs (strain Z). II. Patterns of biosynthesis during the cell cycle. J Cell Comp Physiol 66: 159-182
Edmunds LN Jr (1966) Studies on synchronously dividing cultures of Euglena gracilis Klebs (strain Z). III. Circadian components of cell division. J Cell Physiol 67: 35-44
Edmunds LN Jr (1975) Temporal differentiation in Euglena: circadian phenomena in non-dividing populations and in synchronously dividing cells. In: Les cycles cellulaires et leur blocage chez plusieurs protistes. Centre National de la Recherche Scientifique, Paris (Colloques Intern. du C.N.R.S., No. 240, pp 53-57)
Edmunds LN Jr (1978) Clocked cell cycle clocks: implications toward chronopharmacology and aging. In: Samis HV Jr., Capobianco S (eds) Aging and biological rhythms. Plenum Publishing Corporation, New York (Adv Exp Biol Med, vol 108, pp 125-184)
Edmunds LN Jr (1982) Circadian and infradian rhythms. In: Buetow DE (ed) The biology of Euglena, vol III. Academic Press, New York, pp 53-142
Edmunds LN Jr (ed) (1984) Cell cycle clocks. Marcel Dekker, New York
Edmunds LN Jr (1988) Cellular and molecular bases of biological clocks. Springer-Verlag, New York
Edmunds LN Jr (1994) Clocks, cell cycles, cancer and aging: role of the adenylate cyclase–cyclic AMP–phosphodiesterase axis in signal transduction between circadian oscillator and cell division cycle. In: Pierpaoli W, Regelson W, Fabris N (eds) The aging clock: the pineal gland and other pacemakers in the progression of aging and carcinogenesis. The New York Academy of Sciences, New York (Ann NY Acad Sci, vol 719, pp 77-96)
Edmunds, LN Jr (1995) Regulation of cell division cycles by circadian oscillators: signal transduction between clocks. 'Workshop on 'Towards an Understanding of the Molecular Mechanism of Circadian Rhythms', 10–15 April 1994. Wohldenberg, Germany. Abstracts, no. 33, p 49
Edmunds LN Jr, Adams KJ (1981) Clocked cell cycle clocks. Science 211: 1002-1013
Edmunds LN Jr, Funch RR (1969a) Circadian rhythm of cell division in Euglena: effects of a random illumination regimen. Science 165: 500-503
Edmunds LN Jr, Funch R (1969b) Effects of 'skeleton' photoperiods and high frequency light-dark cycles on the rhythm of cell division in synchronized cultures of Euglena. Planta (Berl) 87: 134-163

Edmunds LN Jr, Halberg F (1981) Circadian time structure of Euglena: a model system amenable to quantification. In: Kaiser HE (ed) Neoplasms - comparative pathology of growth in animals, plants and man. Williams and Wilkins, Baltimore, pp 105-134

Edmunds LN Jr, Laval-Martin DL (1984) Cell division cycles and circadian oscillators. In: Edmunds LN Jr (ed) Cell cycle clocks. Marcel Dekker, New York, pp 295-324

Edmunds LN Jr, Tamponnet C (1990) Oscillator control of cell division cycles in Euglena: role of calcium in circadian timekeeping. In: O'Day DH (ed) Calcium as an intracellular messenger in eucaryotic microbes. American Society for Microbiology, Washington, pp 97-123

Edmunds LN Jr, Jay ME, Kohlmann A, Liu SC, Merriam VH, Sternberg H (1976) The coupling effects of some thiol and other sulfur-containing compounds on the circadian rhythm of cell division in photosynthetic mutants of Euglena. Arch Microbiol 198: 1-8

Edmunds LN Jr, Tay DE, Laval-Martin DL (1982) Cell division cycles and circadian clocks: phase response curve for light perturbations in synchronous cultures of Euglena. Plant Physiol 70: 297-302

Eskin A, Corrent G, Lin C-Y, McAdoo DJ (1982) Mechanism for shifting the phase of a circadian rhythm by serotonin: involvment of cAMP. Proc Nat Acad Sci USA 79: 660-664

Félix M-A, Labbé J-C, Dorée M, Hunt T, Karsenti E (1990) Triggering of cyclin degradation in interphase extracts of amphibian eggs by cdc2 kinase. Nature 346: 379-386

Flitney FW, Singh J (1981) Evidence that cyclic GMP may regulate cyclic AMP metabolism in the isolated frog ventricle. J Mol Cell Cardiol 13: 963-979

Gautier JC, Norbury C, Lohka M, Nurse P, Maller J (1988) Purified maturation-promoting factor contains the product of a Xenopus homolog of the fission yeast cell cycle control gene $cdc2^{+}$. Cell 54: 433-439

Gilman AG (1970) A protein-binding assay for adenosine 3':5'-cyclic monophosphate. Proc Natl Acad Sci USA 67: 305-312

Goldberg ND, Haddox MK (1977) Cyclic GMP metabolism and involvement in biological regulation. Annu Rev Biochem 46: 823-896

Gong J, Traganos F, Darzynkiewicz Z (1993) Simultaneous analysis of cell cycle kinetics at two different DNA ploidy levels are based on DNA content and cyclin B measurements. Cancer Res 53: 5096-5099

Ho IK, Hoskins B (1989) Biochemical methods for neurotoxicological analysis of neuroregulators and cyclic nucleotides. In: Hayes AW (ed) Principles and methods of toxicology. Raven Press, New York, pp 573-598

Hoffmann I, Clarke PR, Marcote MJ, Karsenti E, Draetta G (1993) Phosphorylation and activation of human cdc25-C by cdc2–cyclin B and its involvement in the self-amplification of MPF at mitosis. EMBO J 12: 53-63

Hohmann P, DenHaese G, Greene, RS (1993) Mitotic CDC2 kinase is negatively regulated by cAMP-dependent protein kinase in mouse fibroblast cell free extracts. Cell Prolif 26: 195-204

Homma K, Hastings JW (1989) The S phase is discrete and is controlled by the circadian clock in the marine dinoflagellate Gonyaulax polyedra. Exp Cell Res 182: 635-644

Izumi T, Walker DH, Maller JL (1992) Periodic changes in phosphorylation of the Xenopus cdc25 phosphatase regulate its activity. Molec Biol Cell 3: 927-939

Jarrett RM, Edmunds LN Jr (1970) Persisting circadian rhythm of cell division in a photosynthetic mutant of Euglena. Science 167: 1730-1733

Jessus C, Beach D (1992) Oscillation of MPF is accompanied by periodic association between cdc25 and cdc2-cyclin B. Cell 68: 323-332

John PCL, Sek FJ, Lee MG (1989) A homolog of the cell cycle control protein $p34^{cdc2}$ participates in the division cycle of Chlamydomonas, and a similar protein is detectable in higher plants and remote taxa. Plant Cell 1: 1185-1193

John PCL, Zhang K, Dong C, Diederich L, Wightman F (1993) $p34^{cdc2}$ related proteins in control of cell cycle progression, the switch between division and differentiation in tissue development, and stimulation of division by auxin and cytokinin. Aust J Plant Physiol 20: 503-526

Keirns JJ, Carrit B, Freeman J, Eisenstadt JM, Bitensky MW (1973) Adenosine 3'5' monophosphate in Euglena gracilis. Life Sci 13: 287-302

Klevecz RR (1976) Quantized generation time in mammalian cells as an expression of the cellular clock. Proc Natl Acad Sci USA 73: 4012-4016

Klevecz RR, King GA (1982) Temperature compensation in the mammalian cell cycle. Exp Cell Res 140: 307-313

Krucher NA, Roberts MH (1994) Identification of cdk-like proteins in the eye and brain of the marine snail, Bulla gouldiana. Abstracts, Fourth Meeting of the Society for Research on Biological Rhythms, 4–8 May 1994, Amelia Island, Jacksonville, no. 128, p 96

Lamelli VK (1970) Cleavage of the structural proteins of the head of the bacteriophage T4. Nature 227: 680-685

Ledoigt, G, Calvayrac R (1979) Phénomènes périodiques, métaboliques et structuraux chez un protiste, Euglena gracilis. J Protozool 26: 632-643

Lemmer B, Bissinger H, Lang PH (1986) Effect of forskolin on cAMP levels in rat heart at different times of day. IRCS Med Sci 14: 1103-1104

Lewin B (1990) Driving the cell cycle: M phase kinase, its partners, and substrates. Cell 61: 743-752

Lohka MJ, Hayes MK, Maller JL (1988) Purification of maturation-promoting factor, an intracellular regulator of early mitotic events. Proc Natl Acad Sci USA 85: 3009-3013

Malinowski JR, Laval-Martin DL, Edmunds LN Jr (1985) Circadian oscillators, cell cycles, and singularities: light perturbation of the free-running rhythm of cell division in Euglena. J Comp Physiol B 155: 257-267

Matsumoto K, Uno I, Ishikawa T (1983) Control of cell division in Saccharomyces cerevisiae mutants defective in adenylate cyclase and cAMP-dependent kinase. Exp Cell Res 146: 151-161

McKnight GS (1991) Cyclic AMP second messenger systems. Curr Opin Cell Biol 3: 213-217

Murray AW, Hunt T (1993) The cell cycle: an introduction. W.H. Freeman and Co., New York

Murray AW, Kirschner MW (1989) Cyclin synthesis drives the early embryonic cell cycle. Nature 339: 275-280

Novak, B, Mitchison JM (1986) Changes in CO_2 production in synchronous cultures of the fission yeast Schizosaccharomyces pombe: a periodic cell cycle event that persists after the DNA–division cycle has been blocked. J Cell Sci 86: 191-206

Prosser RA, Gillette MU (1989) The mammalian circadian clock in the suprachiasmatic nuclei is reset in vitro by cAMP. J Neurosci 9: 1073-1081

Roberts MH, Towles JA, Leader NK (1992) Tyrosine kinase regulation of a molluscan circadian clock. Brain Res 592: 170-174

Russell M, Bradshaw-Rouse J, Markwardt D, Heideman W (1993) Changes in gene expression in the Ras/adenylate cyclase system of Saccharomyces cerevisiae: correlation with cAMP levels and growth arrest. Molec. Biol Cell 4: 757-765

Spessert R, Heil K, Gupta BBP, Vollrath L (1992) Biphasic 24-hour variations in cyclic GMP accumulation in the rat pineal gland are due to corrresponding changes in the activity of cytosolic and particulate guanylate cyclase. J Neurochem 59: 2081-2086

Tamponnet C, Edmunds LN Jr (1990) Entrainment and phase-shifting of the circadian rhythm of cell division by calcium in synchronous cultures of the wild-type Z strain and of the ZC achlorophyllous mutant of Euglena gracilis. Plant Physiol 93: 425-431

Thompson WJ, Epstein PM, Strade SJ (1979) Purification and characterization of high affinity cyclic adenosine monophosphate phosphodiesterase from dog kidney. Biochem 18: 5228-5237

Tokiwa G, Tyers M, Volpe T, Futcher B (1994) Inhibition of G1 cyclin activity by the Ras/cAMP pathway in yeast. Nature 371: 342-345

Terry OW, Edmunds LN Jr (1970a) Phasing of cell division by temperature cycles in Euglena cultured autotrophically under continuous illumination. Planta (Berl) 93: 106-127

Terry OW, Edmunds LN Jr (1970b) Rhythmic settling induced by temperature cycles in continuously-stirred autotrophic cultures of Euglena gracilis (Z strain). Planta (Berl) 93: 128-142

Tong J, Edmunds LN Jr (1993) Role of cyclic GMP in the mediation of circadian rhythmicity of the adenylate cyclase–cyclic AMP–phosphodiesterase system in Euglena. Biochem Pharmacol 45: 2087-2091.

Tong J, Carré IA, Edmunds LN Jr (1991) Circadian rhythmicity in the activities of adenylate cyclase and phosphodiesterase in synchronously dividing and stationary-phase cultures of the achlorophyllous ZC mutant of Euglena. J Cell Sci 100: 365-369

Whitfield JF (1990) Calcium, cell cycles, and cancer. CRC Press, Boca Raton, Florida

Whitfield JF, Durkin JP, Franks DJ, Kleine LP, Raptis L, Rixon RH, Sikorska M, Roy Walker P (1987) Calcium, cyclic AMP and protein kinase C-partners in mitogenesis. Cancer Metast Rev 5: 205-250

Yee M-C, Bartholomew JC (1989) Effects of 3-(3,4-dichlorophenyl)-1,1-dimethyl urea on the cell cycle in Euglena gracilis. Plant Physiol 91: 1025-1029

Zatz M (1992) Does the circadian pacemaker act through cyclic AMP to drive the melatonin rhythm in chick pineal cells? J Biol Rhythms 7: 301-311

Zheng X-F, Ruderman JV (1993) Functional analysis of the P box, a domain in cyclin B required for the activation of cdc25. Cell 75: 155-164

6 Membrane Transport and Oscillations in Plants

Fabrice Homblé
Laboratoire de Physiologie Végétale, Université Libre de Bruxelles, Campus Plaine C.P 206/2, B-1050 Brussels, Belgium

Summary

Spatial and temporal oscillations occur in plants. Selected examples are used to show that macroscopic and rhytmic movements of plant organs arise form the activation and regulation of the electrogenic H^+-ATPase and ion channels of the plasma membrane. In some cases, the electrocoupling between H^+-ATPase and voltage-dependent ion channels could provide the basis for a membranar clock. Spatial patterns of pH and electrical potential along plants and plant cells could be a dissipative structure which arises when the H^+-ATPase is working far from equilibrium.

6.1 Introduction

It is now well established that membrane transport processes play a central role in the physiology, growth and development of plants. Indeed, they are involved in cell polarity, signal transduction and propagation, energy production, water transport, solute absorption and translocation, tropic and rhythmic movements of plant organs. All the spatial and temporal oscillations observed in plants imply ion fluxes through the plasma membrane which is the main selective barrier between the cell and its environment. Ion transport through a biological membrane occurs via carriers (e.g., ATPases, PPase, cotransporters) and channels. These integral membrane proteins are ubiquitous in plant cell membranes. As I will show several recent experimental findings show that the regulation of the H^+-ATPase and ion channels of the plasma membrane is of prime importance to form spatial and/or temporal oscillations in plants.

6.2 The Plasma Membrane H^+-ATPase

The H^+-ATPase is the main electrogenic active carrier of the plasma membrane of plants and fungi (Spanswick, 1981; Tazawa and Shimmen, 1982; Slayman, 1987; Serrano, 1988). This enzyme plays crucial roles in various physiological functions such as mineral absorption, phloem loading and unloading, cell elongation, stomatal opening, organ movement etc.. Moreover, different endogenous and exogenous factors such as light, plant growth regulators and phytotoxins are known to regulate directly or indirectly its activity (for review, see Serrano, 1985). The cloning of the gene coding for the plasma membrane H^+-ATPase and the determination of its primary amino acid sequence have provided a preliminary insight into the structure of this enzyme (Serrano, 1988; Harper et al., 1989; Briskin and Hanson, 1992). Clearly, plant plasma membrane H^+-ATPases belong to the P-ATPase family (Pedersen and Carafoli, 1987): they are inhibited by vanadate but not by compounds known to inhibit the F_1F_0-ATPases or the V-ATPases. It has a single catalytic subunit of about 100 kDa and it forms an aspartylphosphate intermediate in the course of its enzymatic reaction cycle (Briskin and Hanson, 1992).

Different approaches have been used to characterize the properties of the plasma membrane H^+-ATPase. Electrophysiological methods were used to investigate the dependence of the current flow through the electrogenic H^+-ATPase upon the electrical membrane potential difference and have provided information about the charge transfer step in the reaction cycle (Gradmann et al. 1978; Hansen et al. 1981; Sanders et al. 1981; Sanders 1990). Purification and reconstitution of the plasma membrane H^+-ATPase were carried out using the biochemical approach. This has permitted to investigate the enzymatic and transport properties of the enzyme (Serrano 1985).

Despite the large amount of data collected these last two decades nothing is known about the molecular mechanism of proton transport through the plasma membrane H^+-ATPase. A complete understanding of this molecular mechanism will depend on detailed information about the three-dimensional structure of the enzyme and about the proton-conducting pathway. The three-dimensional structure of the plasma membrane H^+-ATPase could be obtained from X-ray diffraction if suitable crystals were available: a task which seems particularly tedious with integral membrane proteins. High resolution NMR spectroscopy is also of limited used for this purpose because of the large size of the ATPase polypeptide. On the other hand, the Fourier transform infrared spectroscopy has provided important clues about the secondary structure and molecular mechanism of different membrane integral proteins (Braiman and Rothschild 1988). For instance, the proton conducting pathway of bacteriorhodopsin has been largely elucidated using by time-resolved infrared spectroscopy (Dollinger et al. 1986a, b; Gerwert et al. 1990). FTIR spectroscopy has already been used in plant physiology as a tool to study the composition and structure of the plant cell wall (Homblé et al. 1990; McCann et al. 1992), the structure and function of the photosynthetic reaction centre (Nabedryk et al. 1984; Nabedryck et al. 1990; MacDonald and Barry 1992) and the conformational changes occuring in the Neurospora crassa plasma membrane H^+-ATPase (Goormaghtigh et al. 1994). The secondary structure of the plasma membrane H^+-ATPase is formed from about 40% α-helices, 30% β-sheets, 15% turns and 15 % random coils. The binding of ligands such as MgATP, MgADP and MgVanadate which lock the enzyme in different stages of its catalytic cycle does not change significantly the secondary structure of the H^+-ATPase but modify the tertiary structure of the enzyme (Goormaghtigh et al. 1994)

6.3 Ion Channels

It was early recognized that the electrogenic H^+-ATPase is the primary transport which generates the thermodynamic free energy essential to other solute transports. Therefore, it is generally assumed that membrane transport processes could be regulated by the modulation of the single H^+-ATPase. However, there are recent and numerous evidences showing that ion channels can be subject to a biophysical (e.g., voltage-gated) or a biochemical (ligand-gated) regulation and that the regulation of ion channels is the primary event which triggers some physiological functions.

Ion channels constitute a large group of heterogenous integral membrane proteins. The development of the patch-clamp and channel reconstitution techniques now makes possible analysis of molecular properties of ion channels (Sakmann and Neher 1983, Hedrich and Schroeder 1989; Miller 1986; White and Tester 1994). Ion channels provide a hydrophilic pathway for the diffusion of a particle (a solute molecule or water) through the hydrophobic lipid matrix of the membrane (Homblé and Véry 1992). The flux of particles through an ion channel occurs at a high rate (~10^7 ions/s) down their electrochemical potential gradient (Hille 1984). This high efficiency of transport of ions through ion channels requires only that channels are present in minute quantities, thus, making the biochemical study of ion channel tedious. Ion channel can exist in more than one functional state, and rapid transitions between the states can be regulated by the membrane electrical potential difference, by the binding of a molecule or by the interaction with other membrane proteins like, for instance, specific receptors. Taking into account the molecular properties of ion channels such as ion selectivity, gating characteristics, unitary conductance and pharmacology about twenty different ion channels has been found in plant cells (Tester 1990). Plant ion channels can be regulated by various factors such as the electrical membrane potential difference, membrane mechanical stress, inositol triphosphate, calcium, ATP, photoreceptors, ion concentration, plant growth regulators (plant hormones) (e.g., Hedrich and Neher 1987; Schroeder and Hagiwara 1989; Alexandre et al. 1990; Blatt et al. 1990; Katsuhara et al. 1990; Cosgrove and Hedrich 1991; Lew et al. 1992; Assmann 1993; Blatt and Thiel 1993). The rapid development of molecular biology techniques has opened up novel perspectives for the study of ion channels at the molecular level. cDNAs from Arabidopsis thaliana that suppress the mutant phenotype of potassium transport-deficient strain of Saccharomyces cerevisiae have been isolated (Anderson et al. 1992; Sentenac et al. 1992). DNA sequence analysis of these clones predicts an amino acid sequence with structural similarities to K^+ channels of insects and mamalian cells. These include six putative transmembrane helices clustered in the N-amino half of the protein, the probable voltage-sensor and a pore forming domain. Thus, the elucidation of primary sequences of K^+ channels of plant cells suggests that they belong to a large family not only in terms of functional properties but also in terms of structural homology.

6.4 Temporal Oscillations

Plant oscillations such as leaf movements have fascinated scientists since the 18th century. Today the field of plant and cell movements encompasses several approaches of botany such as plant physiology, anatomy, biophysics, biochemistry and ecology so that a review of this broad field would require more than a single article. Many rhythms have been observed in higher plants. They include leaf movements, petal movements, rates of growth, stomatal opening and closing. However, the use of electrophysiological techniques has enable investigators to address new questions in order to understand the role of plant membrane transport processes in these phenomena.

6.4.1 The Movement of Plant Organs

Despite their relatively slow growth, plants are usually regarded as a static organism. However, some plants have leaves that move in response to endogenous and/or exogenous stimuli. For instance, the leaflets of the Venus flytrap (Dionaea muscipula) fold together within about 300 milliseconds after they have been touched by an insect. The leaves of the sensitive plant (Mimosa pudica) move downwards and its leaflets fold together in a few seconds after the plant is touched. This rapid closure is due at least in part to contractile proteins (Vanden Driessche 1978). In plants showing nyctinastic leaf movements, leaves and leaflets are generally horizontal in the light and fold together in the dark. This circadian rhythm is sustained even in the absence of the day-night alternance which provides an evidence that the movement is controlled by an internal clock. Rhythms with obvious outward manifestation can be caused either by a differential and irreversible growth, or by a differential and reversible transport of water and solutes into or out of specialized cells called motor cells that form a pulvinus.

Pulvini are usually located at the base of the petiole, blade or leaflet. Cells in the pulvinus that swell during opening are called extensors, whereas those that shrink are called flexors. The movements arise from co-ordinated changes in turgor of the motor cells (Satter and Galston 1981). Schildknecht and cowokers (Schlidknecht 1983; Kallas et al. 1990) have found a new class of compounds, called turgorins, that activate pulvini of some nyctinastic plants such as Mimosa, Acacia, Oxalis and Robinia. Most of those periodic leaf movement factors (PLMF) are related to the β-glucosides of gallic acid, with the glucoside bonding at its para-hydroxyl group. Other like the PLMF3 of Oxalis are derivatives of the protocatechuic acid. Those compounds are active at low concentrations (10^{-5} to 10^{-7}M) which suggests that they could play a role in the regulation of the physiology of the pulvinus. Other molecules to intervene are plant growth regulators (at least ethylene, abscisic acid and salicylic acid) (Schrempf 1980; Saeedi et al. 1984; Roblin and Pérault 1985). Several ion channels have been identified in the plasma membrane of Samanea pulvinar motor cells. Moran et al. (1988) have found an outward-rectifying K^+ channel which could account for the large K^+ efflux during the shrinking of the motor cells. If this is true then the membrane of motor cells should depolarize as they shrink.

Not all plants have pulvini. For instance, they are lacking in the Venus flytrap: a carnivorous plant. In this plant, stimulation of leaf hairs (usually by insects and spiders) triggers a series of action potentials which yield the closure of the opposit leaves. In the case of the Venus flytrap it has been proposed that the rapid closure of the leaves is due to an acid stimulated growth triggered by a fast acidification of the extracellular compartment due to the plasma membrane H^+-ATPase (Williams and Bennett 1982). According to the acid-growth hypothesis, the proton pump which is the main electrogenic active transport of plant cells secretes protons into the surrounding wall and lower the local pH so that wall loosening occurs. This allows the wall to stretch more easily which promotes the growth of the cell (Taiz 1984). Several arguments support the hypothesis of Williams and Bennett (1982). Firstly, leaves infiltrate with neutral or slightly acid buffer (>pH 6) do not respond to a stimulus whereas those infitrated with acid buffers (<pH4.75) close spontaneously. Secondly, large amount of ATP are lost during the closure process which agrees with the assumption that the process requires the activation of the plasma membrane proton pump. Finally, there is an irreversible expansion of the outer epidermis during closure and an irreversible expansion of the inner epidermis during the reopening phase.

However, Hodick and Sievers (1989) have questionned the hypothesis. Their data do not reproduce the results of Williams and Bennett (1982): for instance, buffering the cell wall do not prevent movement of traps even at pH 6.

6.4.2 The Action Potential

Many canivorous plants use rapid movements to trap their prey, and in some species it has been shown that the movements are mediated by an action potential viz., a transient depolarization of the membrane similar to that observed in Mimosa seismonastic movements. A rapid movement of the sensitive plant associated with an action potential can be triggered by various stimuli, including light, temperature, electricity and mechanical pressure (Siboaka 1962). The ionic basis of action potential is best charactarized in the large excitable cells of Characeae: an aquatic green plant (Beilby 1979; Lunevsky et al. 1983). The sequence of events which occurs during an action potential can be summarized as follows. In response to an external stimulus there is an initial influx of extracellular calcium which activates calcium-modulated voltage-dependent chloride channels and yields a chloride efflux. Then, the resultant membrane depolarization activates voltage-dependent potassium channels which lead to an efflux of potassium and repolarizes the membrane. Similar ionic fluxes are generally thought to be involved in the action potential of higher plants, although supporting data in the literature are indirect and less detailed (Simon 1981). The plasmalemma ionic currents from the motor cells of the pulvinus of Mimosa were recently investigated by patch-clamp techniques. An outward-rectifying voltage-dependent potassium channel which could account for the membrane repolarisation was characterized in this way (Stoeckel and Takeda 1993). A similar channel was also found in the plama membrane from non excitable pulvinar motor cells of Samanea (Moran et al. 1988; Moran et al. 1990) and from the trap-lobe motor cells of Dionea (Iijima and Hagiwara 1987).

In some cases repetitive and spontaneous action potentials can occur in response to a stimulus including a sudden increase of monovalent salt concentration, removing extracellular calcium by EGTA (Hodick and Sievers 1988) in Dionaea or electrical current (Hayashi and Hirakawa 1980). These oscillations are usually completely damped after an hour or less. Hayashi et al. (1982, 1983) studying Nitella have shown that superposition of a sinusoidal current to this self-sustained oscillation results in entrainment of the spontaneous oscillations as well as quasi periodicity and chaos, depending on the frequency of the sinusoidal signal. The repetitive firing of action potential seems to be triggered when the electrogenicity of the H^+-ATPase is weak (lack of external calcium or membrane current-clamped near the threshold potential for an action potential). Is it a condition to set up an electrocoupling between the membrane transporter in excitable plant cells? We do not know. More experimental data are required to understand the origin of this phenomenon. The magnitude of the action potential decreases with time during the oscillation: the threshold potential becomes more positive and the peak of action potential occurs at more negative electrical potential differences (Radenovic and Pencic 1970). This indicates that ion concentrations change in the cell. In plant cells the H^+-ATPase provides the driving force for solute absorption which requires H^+/solute cotransport. If the H^+-ATPase activity is weak, the intracellular ion concentration will not be reset correctly. This will be particularly true for the chloride that is absorbed via an $2H^+/Cl^-$ symporter (Sanders and Hansen 1981). Therefore, damping of the spontaneous rhythmic activity could arise because plant cells do not have the machinery required in order to keep their cytoplasmic concentrations in steady state during the oscillations. If it is so, the repetitive firing of the action potential must be seen as an epiphenomenon of plant cell physiology.

6.4.3 The Opening-Closing Oscillation of Stomata

The guard cells have an important role for the physiology of the whole plant. They regulate the stomatal aperture in such a way that neither photosynthesis nor respiration nor transpiration becomes a limiting factor for plant growth and development. Changes in pore aperture occurs within a few seconds after the stimulus application. The stomatal opening is an osmotic process. Accumulation of osmotically active solutes drives water uptake which cause the swelling of guard cells. Water and ions (mainly K^+ and Cl^-) are taken up from the apoplast outside the cell (Grignon and Sentenac 1991). The apoplast is formed from the continuum of cell wall. There is an inhomogeneous distribution of the cell wall around the guard cells. The radially distributed cell wall microfibrils of cellulose result in a concave blowing out of the guard cells which increases the pore aperture during water uptake. During stomatal opening, guard cells extrude H^+, take up $K^+$and Cl^-, and produce malate^{2-}. These solutes are stored in the large central vacuole. The H^+ extrusion occurs through the electrogenic H^+-ATPase of the plasma membrane (Assmann and Schwartz 1992). Several factors such as light, phytochrome and plant growth regulators (plant hormones) are known to trigger the activity of the plasma membrane H^+-ATPase (Assmann et al. 1985; Serrano et al. 1988; Lohse and Hedrich 1992). It sets up an electrochemical potential gradient across the plasma membrane. The passive uptake of K^+ down the electrical gradient created by the plasma membrane H^+-ATPase is mediated by inwardly rectifying K^+ channels (Schroeder et al. 1984, 1987; Schroeder 1988; Schroeder and Fang 1991). Although the inward rectifying K^+ channels could account for the K^+ influx measured during the stomatal opening, a stretch-activated K^+ channel could also contributed to the K^+ uptake (Cosgrove and Hedrich 1991). Regulation of the cytoplamic pH is achieved via the production of malate^{2-} (and $2H^+$) which also provides a counter ion for K^+ and rises the intracellular osmotic potential. The K^+ influx occuring during the stomatal opening can also be balanced by an uptake of Cl^- via an H^+/Cl^- symport. However, the molecular caracterization of this putative carrier has not yet been done.

At the molecular level, the stomatal opening-closure is not a reversible process. Indeed, stomatal closure requires specific ion transports. During stomatal closure, there is a passive loss of K^+ which requires a membrane depolarization. The outwardly rectifying K^+ channels provide a pathway for the net efflux of K^+. Those channels are regulated by the voltage and the K^+ concentration so that they are activated under conditions favoring the passive loss of K^+ (Blatt 1988; Hosoi et al. 1988; Schroeder et al. 1984,1987; Schroeder 1988). However, the factor which triggers the membrane depolarization is not known. Before stomatal closure the membrane potential difference can be as high as -160 mV. Activation of the stretch-activated calcium channel of the plasma membrane can occur at membrane potential differences as negative as -180 mV (Cosgrove and Hedrich 1991). The inwardly-rectifying K^+channel which is involved in stomatal opening is also slightly permeable to Ca^{2+} and activated at negative membrane potential differences. There is no experimental evidence showing that the uptake of Ca^{2+} through these channels could be sufficient to depolarize the membrane but one can hypothesize that the local increase of Ca^{2+} concentration could be the signal for a sequence of events which could lead to membrane depolarization. Three other anion-selective channels have been found in the plasma membrane of guard cells. They provide a pathway for the efflux of Cl^- during stomatal closure. The S-type anion channel activates and deactivates slowly in reponse to the voltage (Schroeder and Hagiwara 1989; Schroeder and Keller 1992). The R-type anion channel activates rapidly and then deactivates in response to a membrane depolarization. Moreover, it deactivates rapidly when the membrane is hyperpolarized (Hedrich et al. 1990; Linder and Raschke 1992; Keller et al. 1989; Marten et al. 1991; Schroeder and Keller 1992). A stretch-activated anion channel has also been described (Cosgrove and Hedrich 1991). There is no strong evidence showing that these anion channels could trigger the membrane depolarization because they are activated at voltages more positive than -160 mV. Malate removal during stomatal closure occurs either via an export outside the cell (Van Kirk and Raschke 1978) or via the metabolic pathway (Hedrich et al.1985).

Stomatal movements can be regulated by various endogenous and exogenous factors including cytoplasmic and extracellular pH, plant growth regulators, ion concentration, light, carbon dioxide, humidity and second messengers (Assmann 1993; Blatt and Thiel 1993). Importantly, they integrate a circadian rhythm.

6.4.4 The Membrane Clock

Spontaneous oscillations of the plasma membrane electrical potential difference with periods of 45 s, 15 min and 1 h have been observed in plants (Ogata and Kishimoto 1976, Hansen 1978, Nishizaki 1968, Fisahn et al. 1986). These oscillations are of small amplitude: a few millivolts. Though we do not know what is the origin of these oscillations the membrane response implies that membrane transport processes are involved in these oscillatory phenomenon. The 1 h and 15 min oscillations could be generated by the H^+-ATPases and K^+ channels.

Spontaneous oscillations of the plasma membrane electrical potential difference respectively (Fisahn et al. 1986).

The resting membrane electrical potential difference of plant cells (about -200 mV at an extracellular pH of 7) is much more negative than the diffusion potential and arises from the electrogenic properties of the plasma membrane H^+-ATPase. In plants, light is absorbed by the pigments of chloroplasts in which the photosynthesis occurs and by at least a blue light receptor in the plasma membrane. The activity of the plasma membrane H^+-ATPase is modulated by the light intensity and it has been hypothesized that a product of photosynthesis regulates the plasma membrane ion transport. Hansen (1978) has investigated the frequency response of the electrical membrane potential difference of Nitella when the internodal cell is illuminated by a light whose intensity is modulated sinusoidally. He has shown that a feedback loop exists in the biochemical reactions coupling photosynthesis to the membrane electrical potential difference. The pH could be the cytoplasmic factor involved in this regulation (Tazawa and

Shimmen 1982) but not the ATP concentration which is relatively steady whatever the light conditions. A reaction chain involving a feedback loop may lead to oscillatory phenomena. Therefore, oscillation of the resting membrane electrical potential difference could arise from a metabolic modulation of the H^+-ATPase activity.

Alternatively, oscillation of the membrane electrical potential could be independent of any metabolic reaction. Indeed, an oscillation could arise from the coupling between carriers and channels that are voltage-dependent. As aforesaid in § 6.4.3, the best characterized plant system is guard cells where the major transporters have been investigated with respect to the voltage-dependence and relaxation kinetics of their activities. Under free-running conditions large (about one hundred millivolts) oscillations of the plasma membrane electrical potential difference of guard cells have been recorded (Gradmann et al. 1993). Accounting for the four major ion transporters found in guard cells (the H^+-ATPase, the inward-rectifying K^+ channel, the outward-rectifying K^+ channel, the Cl^- channel) and for a $2H^+/Cl^-$ symporter found in other plant cells (e.g., Sanders and Hansen 1981) Gradmann et al. (1993) have described a theoretical model which simulated these oscillations. The period of these oscillations (about 10 s) are much lower than that of infradian and circadian rhythms of the stomatal movement (e.g., Stalfelt 1963, 1965; Ehrler 1971; Martin and Meidner 1971). Thus, these results suggest that the electro-coupling of ion transport through plant membranes could provide the basis for a stable clock.

6.5 Spatial Oscillations

In addition to temporal oscillations, spatial patterning is also observed in plants. Various species of Characeae develop a banded pattern of alternating acid and alkaline regions along the cell surface when they are exposed to light. The pH of alkaline bands is around 9.5 whereas the pH of the adjacent acid band is about 5.5 (Lucas and Smith 1973). This pH pattern has been found to be associated with a variation in membrane potential along the cell surface. Regions of high electrical potential correspond to acidic zones and those of low electrical potential to alkaline zones. Extracellular currents are found to circulate between the alkaline and the acidic regions in roll-like patterns. The pH banding pattern is also correlated with a spatial distribution of cell wall macromolecules (Homblé et al. 1990). A model of the reaction-diffusion type with nonlinear boundary conditions was proposed to explain the origin of the banded pattern in plants (Toko et al. 1985). It suggests that the local electrical potential could give rise to a bifurcation from a homogeneous state to a stable self-organized state. Oscillation of the electrical pattern occurs during banding formation (Toko et al. 1988). This oscillation could arise from the coupling between photosynthesis and the plasma membrane H^+-ATPase which involves a feedback reaction (see § 6.4.4). Spatial patterns of electric potential and pH also occur near the surface of roots of the azuki bean (Phaseolus chrysanthos) (Iiyama et al. 1985; Toko and

Yamafuji 1988). This is associated with an inhomogenous distribution of cytoplasmic ATP concentration and enzyme activity along the length of the root (Toko et al. 1987). The work done by Toko and coworkers suggests that spatial non uniformities could arise from the coupling between H^+-ATPases and ion channels. These patterns could play an important role in plant growth and development which seem to be controlled by the pattern of electrical potential (Jaffe and Nuccitelli 1977).

6.6 Conclusion

Temporal and spatial oscillations in plants are macroscopic manifestations of the regulation and triggering of various membrane ion transport processes. Most of the plant movements are osmotically driven at the cellular level. The best characterized movement at the molecular level is the stomatal opening-closing circadian rhythm. Both ATPase and ion channels are involved in the movements. The electrocoupling between the H^+-ATPase and voltage-dependent ion channels could provide a membrane oscillator for some plant rhythms. Moreover, it could give rise to spatial patterns which might control the growth and development of plant organs.

Acknowledgments

The author is a Research Associate of the National Fund for Scientific Research which is kindly acknowledged for its financial support.

References

Alexandre J, Lassalles JP, Kado RT (1990) Opening of Ca^{2+} channels in isolated red beet vacuole membrane by inositol 1,4,5-triphosphate. Nature 343: 567-570

Anderson JI, Huprikar SS, Kochian LV, Lucas WJ, Gaber RF (1992) Functional expression of a portable Arabidopsis thaliana potassium channel in Saccharomyces cerevisiae. Proc Natl Acad Sci USA 89: 3736-3740

Assmann SM (1993) Signal transduction in guard cells. Ann Rev Cell Biol 9: 345-375

Assmann SM, Schwartz A (1992) Synergetic effect of light and fusicoccin on stomatal opening: epidermal peel and patch clamp experiments. Plant Physiol 98: 1349-1355

Assmann SM, Simoncini L, Schroeder JI (1985) Blue light activates electrogenic ion pumping in guard cell protoplasts of Vicia faba. Nature 318: 285-287

Beilby MJ, Coster HGL (1979) The action potential in Chara corallina II. Two activation-inactivation transients in voltage clamps of the plasmalemma. Aust J plant Physiol 6: 323-335

Blatt MR (1988) Potassium-dependent bipolar gating of potassium channels in guard cells. J Membrane Biol 102: 235-246

Blatt MR, Thiel G, Trentham DR (1990) Reversible inactivation of K^+ channels of Vicia stomatal guard cells following the photolysis of caged inositol 1,4,5-triphosphate. Nature 346: 766-769

Blatt MR, Thiel G (1993) Hormonal control of ion channel gating. Ann Rev Plant Physiol Plant Mol Biol 44: 543-567

Braiman MS, Rothschild KJ (1988) Fourier transform infrared techniques for probing membrane protein structure. Ann Rev Biophys Biophys Chem 17: 541-570

Briskin DP, Hanson JB (1992) How does the plant plasma membrane H^+-ATPase pump protons. J Exp Bot 43: 269-289

Cosgrove DJ and Hedrich R (1991) Stretch-activated chloride, potassium, and calcium channels coexisting in plasma membranes of guard cells of Vicia faba L. Planta 186: 143-153

Dollinger G, Eisenstein L, Lin SL, Nakanishi K, Termini J (1986a) Fourier transform infrared difference spectroscopy of bacteriorhodopsin and its photoproducts regenerated with deuterated tyrosine. Biochemistry 25: 6524-6533

Dollinger G, Eisenstein L, Lin SL, Nakanishi K, Odashima k, Termini J (1986b) bacteriorhodopsin: Fourier transform infrared difference methods for studies of protonation of carboxyl groups. Methods Enzymol 127: 649-662

Ehrler WL (1971) Periodic noctural stomatal opening of Citrus in steady environment. Physiol Plant 25: 488-492

Fisahn J, Mikschl E, Hansen U-P (1986) Separate oscillations of the electrogenic pump and of a K^+-channel in Nitella as revealed by simultaneous measurement of membrane potential and of resistance. J Exp Bot 37: 34-47

Gerwert K, Souvigner G, Hess B (1990) Simultaneous monitoring of light-induced changes in protein side-group protonation, chromophore isomerization, and backbone motion of bacteriorhodopsin by time-resolved Fourier-transform infrared spectroscopy. Proc Natl Acad Sci USA 87: 9774-9778

Goormaghtigh E, Vigneron L, Scarborough GA, Ruysschaert JM, (1994) Tertiary conformational changes of the Neurospora crassaplasma membrane H^+-ATPase monitored by hydrogen/deuterium exchange kinetics: a Fourier transform infrared spectroscopy approach. J Biol Chem 269: 27409-27413

Gradmann D, Hansen U-P, Long WS, Slayman CL, Warncke J (1978) Current voltage relationships for the plasma membrane and its principal electrogenic pump in Neurospora crassa: I. Steady-stae conditions. In Slayman CL, ed, Current Topics in Membranes and Transport, Volume 16, Academic Press, New York, pp 257-276

Gradmann D, Blatt MR, Thiel G (1993) Electrocoupling of ion transporters in plants. J Membrane Biol 136: 327-332

Grignon C, Sentenac H (1991) pH and ionic conditions in the apoplast. Ann Rev Plant Physiol Plant Mol Biol 42: 103-128

Hansen U-P (1978) Do light-induced changes in the membrane potential of Nitella reflect the feed-back regulation of cytoplasmic parameter? J Membrane Biol 41: 197-224

Hansen U-P, Gradmann D, Sanders D, Slayman CL (1981) Interpretation of current-voltage relationships for "active" ion transport systems: I. Steady-state reaction-kinetic analysis of class-I mechanisms. J Membrane Biol 63: 165-190

Harper JF, Surowy TK, Sussman MR (1989) Molecular cloning and sequence of cDNA encoding the plasma membrane proton pump (H^+-ATPase) of Arabidopsis thaliana. Proc Natl Acad Sci USA 86: 1234-1238

Hayashi H, Hirakawa K (1980) Nitella fluctuation and instability in the membrane potential near threshold. Biophys J 31: 31-44

Hayashi H, Nakao M Hirakawa K (1982) Chaos in the self sustained oscillation of an excitable biological membrane under sinusoidal stimulation. Phys Lett 88A: 265-266

Hayashi H, Nakao M Hirakawa K (1983) Entrained, Harmonic, Quasiperiodic and Chaotic responses of the self-sustained oscillation of Nitella to sinusoidal stimulation. J Phys Soc Jpn 52: 344-351

Hedrich R and Neher E (1987) Cytoplasmic calcium regulates voltage-dependent ion channels in plant vacuoles. Nature 329: 833-836

Hedrich R, Schroeder JI (1989) The physiology of ion channels and electrogenic pumps in higher plants. Ann Rev Plant Physiol 40:539-569

Hedrich R, Bush H, Raschke K (1990) Ca^{2+} and nucleotide dependent regulation of voltage-dependent anion channels in the plasma membrane of guard cells. EMBO J 9: 3889-3892

Hedrich R, Raschke K, Stitt M (1985) A role for fructose 2,6-biphosphate in regulating carbohydrate metabolism in guard cells. Plant Physiol 79: 977-982

Hille B (1984) Ionic channels of excitable membranes. Sinauer Associates, Sunderland

Hodick D, Sievers A (1988) The action potential of Dionaea muscipula Ellis. Planta 174: 8-18

Hodick D, Sievers A (1989) On the mechanism of trap closure of Venus flytrap (Dionaea muscipula Ellis). Planta 179: 32-42

Homblé F,Maiornikoff A, Lannoye R (1990) Correlation between the banding pattern and the cell wall composition in Chara corallina. J Plant Physiol 135: 686-691

Homblé F and Véry AA (1992) Coupling of water and potassium ions in K^+ channels of the tonoplast of Chara. Biophys J 63: 996-999

Hosoi S, Iino M, Shimazaki KI (1988) Outward-rectifying K^+ channels in stomatal guard cell protoplasts. Plant Cell Physiol 29: 907-911

Iijima T, Hagiwara S (1987) Voltage-dependent K^+ channels in protoplasts of trap-lobe cells of Dionaea muscipula

Iiyama S, Toko K, Yamafuji K (1985) Band structure on surface electric potential in growing roots. Biophys Chem 21: 285-293

Jaffe MJ, Nuccitelli R (1977) Electrical controls of development. Ann Rev Biophys Bioengin 6: 445-476

Kallas P, Meier-Augenstein W, Schildknecht H (1990) The structure-activity relationship of the turgorin PLMF1 in the sensitive plant Mimosa pudica L. J Plant Physiol 136: 225-230

Katsuhara M, Mimura T, Tazawa M (1990) ATP-regulated ion channels in the plasma membrane of a Characeae alga, Nitellopsis obtusa. Plant Physiol 93: 343-346

Keller BU, Hedrich R, Raschke K (1989) Voltage-dependent anion channels in the plasma membrane of guard cells. Nature 341: 450-453

Lew RR, Krasnoshtein F, Serlin BS, Schauf CL (1992) Phytochrome activation of K^+ channels and chloroplast rotation in Mougeotia. Plant Physiol 98: 1511-1514

Linder B, Raschke K (1992) A slow anion channel in guard cells, activating at large hyperpolarization, may be principal for stomatal closing. FEBS Lett 313: 27-30

Lohse G, Hedrich R (1992) Characterization of the plasma-membrane H^+-ATPase from Vicia faba guard cells. Planta 188: 206-214

Lucas WJ, Smith FA (1973) The formation of acid and alkaline regions at the surface of Chara corallina cells. J Exp Bot 24: 1-14

Lunevsky VZ,Zherelova OM, Vostrikov IY, Berestovsky GN (1983) Excitation of Characea cell membranes as a result of activation of calcium and chloride channels. J Membrane Biol 72: 43-58

MacDonnald GM, Barry BA (1992) Difference FT-IR study of a novel Biochemical preparation of photosystem II. Biochemistry 31: 9848-9856

Marten I, Lohse G, Hedrich R (1991) Plant growth hormones control voltage-dependent activity of anion channels in plasma membranes of guard cells. Nature 353: 758-762

Martin ES, Meidner H (1972) The phase-response of the dark stomatal rhythm in Tradescantia virginiana to light and dark treatments. New Phytol 71: 1045-1057

McCann MC, Hammouri M, Wilson R, Belton P, Roberts K (1992) Fourier transform infrared microspectroscopy is a new way to look at plant cell walls. Plant Physiol 100: 1940-1947

Miller C (1986) Ion channel reconstitution. Plenum Press, New York

Moran N, Ehrenstein G, Isawa K, Mischke C,Bare C, Satter RL (1988) Potassium channels in motor cells of Samanea saman. Plant Physiol 88: 643-648

Moran N, Fox D, Satter RL (1990) Interaction of the depolarization-activated K^+ channels in Samanea saman with inorganic ions. Plant Physiol 94: 424-431

Nabedryk E, Andrianambinintsoa S, Breton J (1984) Transmembrane orientation of α-helices in the thylakoid membrane and in the light-harvesting complex. Biochim Biophys Acta 765: 380-387

Nabedryk E, Andrianambinintsoa S, Berger G, Leonhard M, Mäntele W, Breton J (1990) Characterization of bonding interactions of the intermediary electron acceptor in the reaction center of photosystem II by FTIR spectroscopy. Biochim Biophys Acta 1016: 49-54

Nishizaki Y (1968) Rhythmic changes in the resting potential of a single plant cell. Plant Cell Physiol 9: 613-616

Ogata K, Kishimoto U (1976) Rhythmic change of membrane potential and cyclosis of Nitella internode. Plant Cell Physiol 17: 201-207

Pedersen PL, Carafoli E (1987) Ion motive ATPases. Part I. Ubiquity, properties and significance to cell function. Trends Biochem Sci 12: 146-150

Radenovic C, Pencic M (1970) Oscillation bioélectriques dans des membranes de Nitella. Physiol Plant 23: 697-703

Roblin G, Pérault J-M (1985) Effect of Ethephon, 1-aminocyclopropane-1-carboxylic acid, and inhibitors of ethylene synthesis on the gravitropically induced movement of Mimosa pudica pulvinus. Plant Physiol 77: 922-925

Saeedi S, Gaillochet J, Bonmort J, Roblin G (1984) Effect of salicylic and acetylsalicylic acids on the scotonastic and photonastic leaflet movements of Cassia fasciculata. Plant Physiol 76: 851-853

Sakmann B and Neher E (1983) Single-channel Recording. Plenum, New York

Sanders D, Hansen U-P (1981) Mechanism of Cl^- transport at the plasma membrane of Chara corallina: Transinhibition and the determination of H^+/Cl^- binding order from a reaction kinetic model. J Membrane Biol 58: 139-153

Sanders D, Hansen U-P, Slayman CL (1981) Role of the plasma membrane proton pump in pH regulation in non-animal cells. Proc Natl Acad Sci USA 78: 5903-5907

Sanders D (1990) Kinetic modeling of plant and fungal membrane transport systems. Ann Rev Plant Physiol Plant Molec Biol 41: 77-107

Satter RL, Galston AW (1981) Mechanism of control of leaf movements. Ann Rev Plant Physiol. 32: 83-110

Schrempf M (1980) The action of abscisic acid on the circadian petal movement of Kalanchoe blossfeldiana. Z Pflanzenphysiol 100: 397-407

Schroeder JI (1988) K^+ transport properties of K^+ channels in the plasma membrane of Vicia faba guard cells. J Gen Physiol 92: 667-683

Schroeder JI, Fang HH (1991) Inward rectifying K^+ channels in guard cells provide a mechanism for low-affinity K^+ uptake. Proc Natl Acad Sci USA. 88:11583-11587

Schroeder JI, Hagiwara S (1989) Cytosolic calcium regulates ion channels in the plasma membrane of Vicia faba guard cells. Nature 338: 427-430

Schroeder JI, Keller BU (1992) Two types of anion channel currents in guard cells with distinct voltage regulation. Proc Natl Acad Sci USA 89: 5025-5029
Schroeder JI, Hedrich R, Fernandez JM (1994) Potassium selective single channels in guard cell protoplasts of Vicia faba. Nature 312: 361-362
Schroeder JI, Raschke K, Neher E (1987) Voltage dependence of K^+ channels in guard cell protoplasts. Proc Natl Acad Sci USA 84: 4108-4112
Schildknecht H (1983) Turgorins, hormones of the endogenous daily rhythms of higher organized plants. Angew Chem Int Ed Engl 22: 695-710
Sentenac H, Bonneaud N, Minet N, Lacroute M, Salmon F, Gaymard J-F, Grignon C (1992) Cloning and expression in yeast of a potassium ion transport system. Science 256: 663-665
Serrano EE, Zeiger E, Hagiwara S (1988) Red light stimulates an electrogenic ion pump in Vicia guard cell protoplasts. Proc Natl Acad Sci USA 85: 436-440
Serrano R (1985) Plasma membrane ATPase of plant and fungi. CRC Press, Boca Raton
Serrano R (1988) Structure and function of proton translocating ATPase in plasma membranes of plants and fungi. Biochim Biophys Acta 947: 1-28
Siboaka T (1962) Excitable cells in Mimosa. Science 137: 236
Simon PJ (1981) The role of electricity in plant movements. New Phytol 87: 11-37
Slayman CL (1987) The plasma membrane ATPase of Neurospora: a proton-pumping electroenzyme. J Bioenerg Biomembr 19: 1-20
Spanswick RM (1981) Electrogenic ion pumps. Ann Rev Plant Phys 32: 267-289
Stalfeld MG (1963) Diurnal dark reactions in the stomatal movements. Physiol Plant 16: 756-766
Stalfeld MG (1965) The relation between the endogenous and induced elements of the stomatal movements. Physiol Plant 18: 177-184
Stoeckel H, Takeda K (1993) Plasmalemmal, voltage-dependent ionic currents from excitable pulvinar motor cells of Mimosa pudica. J Membrane Biol 131: 179-192
Taiz L (1984) Plant cell expansion: regulation of cell wall mechanical properties. Ann Rev Plant Physiol 35: 585-622
Tazawa M, Shimmen T (1982) Control of electrogenesis by ATP, Mg^{2+}, H^+, and light in perfused cells of Chara. In Slayman CL, ed, Current Topics in Membranes and Transport, Volume 16, Academic Press, New York, pp 49-67
Tester M (1990) Plant ion channels: whole-cell and single-channel studies. New Phytol 114: 305-340
Toko K, Yamafuji K (1988) Spontaneous formation of the spatial pattern of electric potential in biological systems. Ferroelectrics 86: 269-279
Toko K, Chosa H, Yamafuji K (1985) Dissipative structure in the Characeae: spatial pattern of proton flux as a dissipative structure in Characeae. J Theor Biol 114: 127-175
Toko K, Iiyama S Tanaka C, Hayashi K, Yamafuji K, Yamafuji K (1987) Relation of growth process to spatial patterns of electric potential and enzyme activity in bean roots. Biophys Chem 27: 39-58
Vanden Driessche T (1978) The molecular mechanism of Mimosa leaf seismonastic movement. A re-evaluation. Arch Biol 89: 435-449
Van Kirk CA, Raschke K (1978) Release of malate from epidermal strip during stomatal closure. Plant Physiol 61:474-475
White PJ, Tester M (1994) Using planar lipid-bilayers to study plant ion channels. Physiol Plant 91:770-774
Williams SE, Bennett AB (1982). Leaf closure in the Venus flytrap: an acid growth response. Science 218: 1120-1121

7 Circadian Rhythms and Photoperception in Plants: The Role of Red Light and Blue Light

Han Asard and Roland J. Caubergs
Department of Biology, University of Antwerp (RUCA), Groenenborgerlaan 171, B-2020 Antwerp, Belgium

Summary

The induction of rhythmic phenomena in higher plants, algae and fungi clearly involves non-photosynthetic pigments. The construction of action spectra and the red/far red reversibility test have proven the involvement of phytochrome and a (at least one) blue/UV photoreceptor. Both photoreceptor systems are involved in a variety of photomorphogenetic and photoperiodic responses.

In recent years our understanding on the nature of these receptors and their potential mechanisms of action have significantly advanced. For example a detailed molecular structure of different types of phytochrome is now available. On the basis of this knowledge molecular models have been constructed to explain the involvement of phytochrome in a diversity of circadian responses. The isolation of phytochrome deficient mutants has recently led to the identification of G-proteins, the Ca^{2+}/calmodulin system and cGMP as possible intermediates in the signal transduction pathway of this photoreceptor.

At the level of blue light perception it becomes increasingly evident that the heterogeneity of physiological responses reflects the existence of different and interacting photoreceptor molecules. Flavins, carotenoids and pterins have been suggested to be involved in a number of blue light-inducible phenomena. In the mean time progress has been made towards the understanding of blue light effects at the plasma membrane level. The rapid blue light-induced phosphorylation of a specific plasma membrane protein likely indicates an early signal transduction reaction. Also oxidation-reduction phenomena at the plasma membrane, involving blue light-sensitive flavins and a *b*-type cytochrome are now characterized in more detail.

7.1 Introduction: plants and light (with exclusion of photo-synthesis)

The major energy source of this planet, light, is available to the living creatures only in daily cycles. It is therefore not surprising that many vital processes in cells and organisms show similar circadian rhythmicity (CR). It should also be no surprise that the perception and transduction of rhythmic signals from the environment is closely related to the energy status of the cells (Wagner et al. this volume). This energy status is largely defined by the intracellular levels of ATP and reducing equivalents and by electrochemical gradients across the membranes. Also, many intracellular processes are regulated through the functions of membranes and transport across lipid bilayers. It is therefore likely that membranes may be an important regulatory site in the generation and transduction of rhythmicity.

Although several physical parameters may be involved in the synchronization and entraining of biological phenomena, light is the most important 'Zeitgeber'. Not only is light energy a precious source that should be captured optimally, it is also a source of information on the rhythmic cycles for the organisms living on this planet. Photoperiodic mechanisms affect many developmental processes including flowering. Length of day measurements lie at the basis of the induction of these phenomena. Although many cell metabolic activities show circadian fluctuations, and may be part of the oscillatory mechanism, some components sensitive to light must initially perceive signals.

In recent years our knowledge on the nature and distribution of non-chlorophyll plant photoreceptors has significantly advanced (Furuya 1993, Vierstra 1993, Short and Briggs 1994, Chory 1994). This has also led to new hypotheses on their mechanism of action. In particular the discovery of several isoforms of the photoreversible photoreceptor phytochrome opened new perspectives on the operation of this pigment. It is known for many years that phytochrome plays an essential role in photoperiodic induction and this role has been reviewed recently (Lumsden 1991, Thomas 1991). As pointed out by Raven (1983) photoreceptors of photomorphogenetic responses such as phytochrome a blue light receptor are likely to act notably through the regulation of solute fluxes, possibly at the plasma membrane. This chapter therefore notably emphasizes on the evidences for photoreceptor actions at the level of this organelle. Blue and red light signals have been demonstrated to affect gene transcription and gene translation reactions in several plants. These subjects are reviewed recently (Quail 1991, Furuya 1993, Short and Briggs 1993) and will not be covered in detail here.

7.2 Phytochrome

7.2.1 The role of phytochrome

Several physiological responses involving photoperiodic induction have clearly demonstrated the role of phytochrome. Well studied phenomena include the rhythmic leaflet movement of *Samanea* or *Albizzia* and flowering responses. The interaction of phytochrome with many of these CR has recently been reviewed by Lumsden (1991) and Thomas (1991).

As with most other phytochrome-controlled phenomena in plants, the underlying action mechanisms have not yet been elucidated. Phytochrome-mediated reactions have been described at several levels of cell organization including gene expression, ion transport and several metabolic activities (Quail 1991, Furuya 1993, Lumsden 1991). Several hypotheses on the action mechanism for phytochrome have included changes in cellular calcium concentrations as a second messenger (e.g. Haupt 1982). However, only recently direct biochemical evidence was presented supporting these ideas. Using the phytochrome-deficient tomato mutant *aureus* in a single cell assay system, Neuhaus et al. (1993) and Bowler et al. (1994) presented strong evidence for the involvement of G-proteins, calcium ions and cGMP in phytochrome-controlled phenomena. Apparently chloroplast development and anthocyanin synthesis shared the dependence on the activation of G-proteins in the phytochrome signal transduction pathway. The main implication of these experiments is probably that they support a signalling mechanism that is very similar in several aspects to well known animal signal transduction pathways. This recent hypothesis however does not (yet) provide an explanation for the potential role of membrane transport processes as deduced from many physiological experiments.

Two types of phytochrome have been identified in many plants for several years now (for review see Furuya 1993). The so called Type I phytochrome (equivalent to phytochrome A) is predominantly found in large quantities in etiolated tissues and is suggested to rapidly degrade in most species upon transition to light grown conditions. The other phytochrome type (Type II) includes several isoforms (phytochromes B-E) and is of major importance in green tissues. The genes for the apoproteins of these different phytochrome molecules have been sequenced form *Arabidopsis*, they show maximal sequence homologies between 50-80% and the protein synthesis is regulated by feedback control (Sharrock and Quail 1989, Pratt 1995).

On the basis of physiological evidence two pools of phytochrome, with different stabilities in the activated Pfr form, have been suggested to operate in the photoperiodic regulation in short day plants (Lumsden et al. 1987, Thomas 1991). However, the existence of a Type I and Type II phytochrome molecule in both etiolated and green plants does not seem to provide a simple molecular answer for the induction of photoperiodism. *In vivo* phytochrome molecules combine into Y-shaped dimeric structures with two chromophores (Edgerton and Jones 1992). Therefore the existence of heterodimers (Pr:Pfr) could provide an explanation for certain inductive processes (Thomas 1991).

7.2.2 Phytochrome, membranes and CR

Phytochrome responses have often been associated with direct action of the photoreceptor at some subcellular membrane fraction. For example several experiments, in a variety of organisms, point to alterations in solute fluxes apparently controlled by phytochrome activation. Red light-induced calcium fluxes were observed in oat coleoptile cells (Halle and Roux 1980), in isolated protoplasts (Bossen et al. 1988, Takagi and Nagai 1988) and *Vallisneria* mesophyll cells (Takagi et al. 1990). The involvement of phytochrome in the transport of nitrate and ammonium in the green alga *Ulva rigida* was also proposed (Lopez-Figueroa and Rüdiger 1991). Despite considerable efforts however, the hypothesis on the membrane action of phytochrome is still a matter of speculation. The main problem is the difficulty to unambiguously demonstrate the biochemical association of phytochrome with some membrane fraction. In fact, subcellular fractionation studies and immunological techniques indicated the presence of active phytochrome in every fraction tested (Manabe and Furuya 1975, Pratt 1979). Microbeam irradiation techniques suggested the involvement of phytochrome located near the plasma membrane in the particular cases of chloroplast movement in *Mougeotia* (Haupt 1982) and the growth of fern protonemata (Furuya et al. 1980). However, these results should be taken with caution because they do not exclude the action of cytoplasmic or cytoskeleton associated phytochrome molecules.

The possible action of phytochrome at a well defined subcellular fraction has been traced back to the so called 'pelletability' and phytochrome sequestering (Furuya 1993) phenomenon. Red light irradiation caused a fraction of phytochrome molecules to associate with a crude pelletable membrane fraction and it induced an *in situ* redistribution of phytochrome molecules in discrete small particles. Although both effects are far red light reversible they did not lead to a further understanding of the possible phytochrome action at the membrane level. Pelletability has now apparently been explained in terms of electrostatic interactions between the photoreceptor molecule and membrane phospholipids (Furuya et al. 1981). This association is ascribed to Type I phytochrome and might turn out to be of little physiological importance. Thus, although many researchers shared the intuitive conviction that phytochrome directly acts at the membrane level, no solid evidence is yet available.

Despite these difficulties, some recent lines of evidence have still appeared and give new support to the 'membrane hypothesis' for phytochrome action. First, a small fraction of phytochrome (0.5 to 10 % of total) in etiolated tissues was consistently recovered with a membrane fraction and showed distinct hydrophobic properties as demonstrated by partitioning in the detergent Triton X-114 (Lamparter et al. 1992). To explain these results an association between phytochrome and some membrane protein which is not an unspecific binding to lipids was proposed. The nature of the membrane fraction and binding protein however remain to be resolved. A recent analysis of photochemical and fluorescence properties provides additional evidence for the existence of membrane-bound phytochrome in the cell (Sineshchekov et al. 1994). Second, recent immunolocalization studies on the green alga *Mougeotia* suggest the presence of a small fraction of active (Pfr) phytochrome bound to a reactant at the plasma membrane and in equilibrium with the cytoplasmic phytochrome pool (Hanstein et al. 1991). A final and, although indirect argument but strongly appealing was recently formulated in a review paper by Chory (1994). The convincing demonstration of the involvement of trimeric G-proteins in the early signal transduction steps for phytochrome (Neuhaus et al. 1993, Bowler et al. 1994) implicates the probable involvement of transmembrane receptor proteins. Also the remarkable similarities between rhodopsin and phytochrome signalling are suggestive in this respect.

Thus, although the idea of phytochrome acting at a membrane site does not feature in most reviews any more, recent results may renew the interest in this hypothesis. Also the involvement of G-proteins may prove of great importance in the understanding of the role of phytochrome in CR-inductive reactions.

7.3 Blue light

7.3.1 Blue light and CR

A large variety of light-dependent processes in a wide range of organisms show action spectra with a dominant maximum between 300 and 500 nm. The so called blue light-phenomena are numerous and well known and include phototropism, growth responses, phototaxis, enzyme regulation and gene expression (for reviews see Senger 1984, 1987). Blue light photoperception is of particular importance in algae and fungi since these organisms generally lack phytochrome. In algae and fungi blue light is absolutely required for photomorphogenetic responses and for CR synchronization (Schmid 1987, Dring 1988, see also Vanden Driessche this volume). For example *Neurospora crassa* conidiation is synchronized by blue light (Feldman and Dunlap 1983) as well as the phototactic behaviour of *Chlamydomonas* cells (Kondo et al. 1988). In *Acetabularia*, CR synchronization (as well as morphogenetic induction) depends on blue light only whereas the entraining light spectrum is more complex for *Gonyaulax* (see this review for references). A review on CR's in the unicellular organisms *Gonyaulax*, *Euglena* and *Acetabularia* has recently appeared (Vanden Driessche 1994). Blue light is also involved in the induction of rhythmic leaflet movements in *Leguminosea* (Satter and Galston 1981).

7.3.2 Nature of the blue light photoreceptor(s)

The discussion on the chemical nature of the photoreceptor for blue light phenomena generally focuses on the discrimination between flavins and carotenoids. Each of these molecules shows absorption spectra that closely resemble the action spectra of several physiological responses. Arguments in favour of either of these molecules, based on photochemical properties and action spectrum details, have been presented (Song 1984). Moreover, the possible photoperception by pterins in certain blue light-mediated physiological processes has been suggested (Galland 1988). Pterin molecules are composed of a bicyclic ring system, the pyrazone [2,3-d]pyrimidine or pteridine moiety and their biochemical and physiological importance is known from work with bacteria, insects and vertebrates. Pterin-derived pigments possess photochemical and redox properties that make them suitable as photoreceptors.

A major draw back in the chemical analysis of the potential blue light photoreceptor molecule is its apparent very low concentration in the cell. In addition there is no easy photochemical assay available comparable to the red/far red light reversibility test of phytochrome that could unequivocally identify the blue light photoreceptor molecule in *in vivo* and *in vitro* studies. A model has been proposed for blue light photoperception in *Commelina* leaves involving the conversion of a blue light sensitive molecule to an active state (light induced) and the reversion of that molecule to the inactive state (driven by light or thermal energy) (Iino et al. 1985). These photochemical and physical properties have not resulted however in an easy to use detection system. Also a particular blue light response detected in isolated higher plant membrane fractions (LIAC activity, see below) has not yet proven useful in the identification of a blue light photoreceptor, for example in protein purification experiments.

The fact that mutant organisms are relatively easily generated in many fungal species, resulted in several well defined mutants with defective or strongly reduced blue light responses (e.g. Galland and Lipson 1984). These mutants include *Neurospora* strains defective in conidiation CR (Russo 1988). One particular *Neurospora* mutant strain was found to be auxotroph for riboflavin. In this mutant a good correlation was found between the concentration of riboflavin in the culture medium and the sensitivity to blue light-mediated phase shifting (Paietta and Sargent 1983). In addition the use of green-sensitive flavin analogs resulted in similar shifts of the action spectrum. In this particular case the use of mutants has resulted in strong evidence for the involvement of flavins as a blue light photoreceptor.

Although several of the mutations in *Neurospora* and *Phycomyces* were postulated to be affected in the early signal transduction or photoperception, they have not yet resulted in the identification and isolation of a photoreceptor molecule. The availability of blue light defective mutants in *Arabidopsis* on the other hand has recently led to the isolation of a gene (*HY4*) that shows strong homologies to microbial photolyases involved in light-mediated DNA-repair (Ahmad and Cashmore 1993). The *HY4* gene product is likely to bind flavin derivatives and could therefore possibly act as a blue light photoreceptor. Up to now there is no indication on the mode of action or subcellular localization of the *HY4* gene product. Photolyases are also known to contain pterin derivatives as coreceptors and as a result the nature of the chromophore of this protein is still uncertain.

Physiological and genetic evidence has so far pointed to flavins and carotenoids as the main photoreceptor candidates in blue light photoperception. Studies on blue light-induced phenomena have also resulted in the detection of a number of effects in cell free systems. Because of the largely reduced level of complexity of these systems, it was expected that the chemical nature of the blue light receptors involved in these reactions could be more easily addressed. Blue light reactions in subcellular fractions include enhanced protein phosphorylation, light-mediated cytochrome reduction (LIAC) and G-protein activation (see section 7.4.2). Photochemical evidence and the use of specific flavin quenchers have pointed to the involvement of flavins in each of these phenomena.

LIAC activity reflects the blue light-induced reduction of a specific *b*-type cytochrome. Action spectroscopy and the effect of specific inhibitors such as KI and phenylacetic acid have suggested the involvement of a flavin chromophore in this reaction (Goldsmith et al. 1981, Asard and Caubergs 1991). A particular property of the LIAC activity however is its high sensitivity to the addition of exogenous flavins (Goldsmith et al. 1981, Asard et al. 1995). This feature has made it particularly difficult to rule out the effect of endogenous free flavins that could associate with plasma membrane components in a non-specific manner. Recent evidence from our laboratory (Asard et al. 1995) indicates that endogenous redox components, including flavins, generate conditions sufficient to allow blue light-mediated cytochrome *b* reduction.

Recent experiments pointed to the presence of blue light-stimulated GTPase activity at the plasma membrane from etiolated pea tissues (Warpeha et al. 1991). The presence of G-proteins in plant plasma membranes has already been demonstrated in several plant systems (for reviews see Terryn et al. 1993 and Kaufman 1994) and was confirmed by Western blot analysis in the work of Warpeha. Fluorescence analysis indicated the possible action of a flavoprotein in this particular response (Warpeha et al. 1992). Also flavin triplet state quenchers were effective in inhibiting the GTPase activation. Although the action of carotenoid chromophores was ruled out by these experiments, they were unable to discriminate between flavins and pterin-like molecules.

Kinetic and photochemical properties of the blue light-mediated phosphorylation reaction have been studied and reviewed by Short and Briggs (1994). Supression of the protein phosphorylation by inhibitors of flavin excitation suggest the participation of a flavoprotein. The authors discuss several lines of evidence that indicate that the photoreceptor, the protein kinase activity and substrate protein may be closely linked to each other or may even be combined in a single functional protein. The structure of the 120 kDa phosphorylated protein is currently under investigation and may soon lead to more conclusive answers on the nature of the photoreceptor.

7.3.3 Evidence for several blue light receptors

Although the discussion on flavins versus carotenoids and including the possible role of pterins is not yet settled, the involvement of flavins in most blue light-mediated phenomena is generally accepted. It would however be an oversimplification to think of flavins as 'The' blue light receptor. Several lines of evidence clearly indicate the action of different pigments in certain responses. In fact, a close investigation of the action spectra reveals consistent differences that can only be ascribed to the presence of different chromophores (Briggs and Iino 1983). From this analysis at least three distinct classes of blue light photoreceptors were distinguished. The first group of action spectra most strongly suggested the involvement of flavins. Photoresponses ascribed to these receptors include phototropism of etiolated coleoptiles and sporangiophores, light-induced carotenogenesis in *Fusarium* and suppression of the conidiation CR in *Neurospora*. The flavin-mediated cytochrome reduction (LIAC) has long been associated with this class of blue light reactions. The second class of action spectra could be attributed to the action of either a flavin or carotenoid chromophore. Several lines of evidence for example suggest that light-induced carotenoid synthesis in *Neurospora* is affected by a different chromophore than the conidiation suppression response (Schmid 1987). In a third class of action spectra, with a sharp maximum near 470 nm, neither a flavin or a carotene-like pigment could be held as a promising candidate.

There is also evidence that several interacting chromophores may be involved in the blue light perception of a single physiological response. Fluence-response curves for phototropism of several higher plants, including *Zea*, *Avena*, *Arabidopsis*, *Cucumis* and *Raphanus*, and of the sporangiophore of *Phycomyces* show complex kinetics (Galland and Lipson 1987, Galland 1990, Konjevic et al. 1989). These results have been interpreted as the action of different photoreceptors or different chromophores on a single protein. In addition genetic evidence supports the existence of separate photoreceptors in the growth inhibition response of *Arabidopsis* hypocotyls (Konjevic et al. 1992). Finally the picture on blue light photoperception in higher plants is further complicated by the recent results from Quinones and Zeiger (1994), suggesting the involvement of the xanthophyll, zeaxanthin in photoresponses of corn coleoptiles.

At the level of the blue light responses in subcellular fractions (see above) different sensitivities are found towards flavin inhibitors. For example, G-protein activation was inhibited by KI at approximately a 100-fold lower concentration than blue light-induced protein phosphorylation (Short and Briggs 1994). These results clearly support the involvement of different chromophores or alternatively different protein and/or membrane environments of the respective flavins. Finally a further level of complexity in blue light induced phenomena arises from the possible interaction of phytochrome in certain blue light responses of higher plants.

Thus, in conclusion, although flavins are apparently involved as photoreceptors in many blue light phenomena, it is clear that other photoreceptors may participate in certain responses. Caution should therefore be taken in referring to a single blue light photoreceptor for blue light physiological responses. Also the term cryptochrome, suggesting a unique receptor is misleading in that respect (Senger 1984).

7.4 Blue light perception and the plasma membrane

7.4.1 Evidence for the photoreceptor localization

Blue light phenomena in higher plants and in lower organisms are often associated with a directional response (e.g. tropic and tactic responses). It is therefore necessary that any signal perception system should in some way transfer the light information onto a solid subcellular structure, located in the cell and capable of transforming the directional information into a differential response. This reasoning has been an important argument suggesting the plasma membrane as a site of blue light photoperception. The spatial distribution-argument for the photoreceptor localization at the plasma membrane has been most elegantly worked out in a mathematical analysis using the sporangiophore of *Phycomyces* as a model system (Steinhardt et al. 1989, Fukshansky and Richter 1990).

A common property to several blue light-mediated phenomena includes changes in ion transport and membrane potential properties of the illuminated cells. For example membrane potential oscillations are suggested to drive K^+-transport in *Samanea* pulvini (Antkoviak et al. 1991). In addition protoplasts from extensor and flexor cells showed differential behaviour of K^+ channels upon blue light-induced hyperpolarization (Kim et al. 1992). Also guard cell protoplasts from *Vicia* have been used in patch clamp experiments to demonstrate transient hyperpolarization upon blue light pulses (Assman et al. 1985, Shimazaki et al. 1986). Blue light affects the uptake of nitrate and nitrite in the green alga *Monoraphidium braunii* (Aparicio and Quinones 1991); electrical potential differences in the filamentous gametophyte of the fern *Onoclea sensibilis* are induced by blue light (Racusen and Cooke 1982). All of these responses point to early effects of the blue light signal at the level of the plasma membrane.

Evidence for the plasma membrane localization of the blue light sensitive pigment(s) also comes from the only available *in vitro* blue light effects that point to the possible action of physiological photoreceptors. Each of the activities of protein phosphorylation, cytochrome *b* reduction and GTPase activation has been located at the plasma membrane. For example light-mediated cytochrome *b* reduction cosedimented with enzymatic markers for the plasma membrane on sucrose density gradients (Asard et al. 1987). The availability of highly purified plasma membrane preparations from several plant tissues through aqueous polymer two-phase partitioning confirmed this membrane location. Blue light-induced protein phosphorylation and activation of G-proteins was demonstrated in plasma membrane enriched fractions from several species (Short and Briggs 1994).

Finally, it is a necessary prerequisite that the photoreceptor chromophores that have been suggested to operate in blue light signal perception are also found at the plasma membrane. Flavins have indeed been detected in isolated plasma membrane fractions from several species (Leong et al. 1981, Ramirez et al. 1984, Caubergs, unpublished). Apart from their possible role in blue light perception, flavins are likely to be involved in NAD(P)H-dependent plasma membrane oxidoreductase activities (Rubinstein and Luster 1993, Berczi and Asard 1994). So far little qualitative or guantitative data are available on these flavins from the plant plasma membrane. At least a small fraction is apparently tightly bound to proteins and only released upon protein digestion (Caubergs unpublished). Recently we were also able to demonstrate the presence of pterins in highly purified plasma membrane fractions from etiolated *Phaseolus* hypocotyls. Therefore, the only molecules that may be involved in blue light perception that have not been detected in plasma membranes so far are carotenes. This apparently indicates that blue light photoperception mediated by these carotene pigments operates through mechanisms not involving the plasmalemma and probably not involving transmembrane ion transport.

7.4.2 Early steps in blue light transduction

The difficulties in the unambiguous identification of a photoreceptor molecule *in vivo* and *in vitro* have proven to be a major obstacle in the analysis of blue light signal transduction. Among the multitude of physiological effects attributed to blue light only a few have shed some light on the possible molecular actions involved.

One of the most promising recent developments is the discovery by Gallagher et al. (1988) and Short and Briggs (1990) of the phosphorylation of a specific 120 kDa protein in etiolated pea stems. This protein was located at the plasma membrane and homologous proteins of slightly varying molecular masses have been found in several other plant species (Reymond et al. 1992). The phosphorylated protein was not lost from the isolated membranes virtually eliminating the possibility of a non-specific membrane association. Considering the potential role of protein phosphorylation in many regulatory and signal transduction processes in plants (for review see Ranjeva and Boudet 1987), this reaction pointed to a possible mechanism for blue light signal transduction. A detailed analysis of tissue distribution, reaction kinetics, fluence-response relationships and action spectra suggested a close correlation between blue light-mediated protein phosphorylation and phototropism in pea and maize (Short and Briggs 1990, Palmer et al. 1993, Hager and Brich 1993). This correlative evidence has recently also been supported by genetic evidence. Reymond et al. (1992) have shown that an *Arabidopsis thaliana* mutant strain with a highly reduced phototropic sensitivity (JK224) also showed a reduced photophosphorylation response. Two papers add a circadian dimension to protein phosphorylation. The first reported circadian changes in protein phosphorylation in *Gonyaulax* (Schröder-Lorenz and Rensing 1987). In the same organism, the second paper shows that inhibition of protein phosphorylation stops the circadian oscillator and blocks light-induced phase shifts (Comolli et al. 1994).

Observations made by Hager et al. (1993) on the blue light-mediated phosphorylation are of particular interest to plasma membrane redox reactions. Hager et al. demonstrated that potential electron donors such as NAD(P)H, ascorbate and glutathione enhanced the increase in phosphorylation. Oxidoreductases utilizing these molecules as electron donors have been demonstrated at the plasma membrane of many higher plants (Rubinstein and Luster 1993, Berczi and Asard 1994). Some of these enzymes have been purified and turn out also to contain flavin chromophores (Berczi and Asard 1994, Caubergs unpublished).

The recent discoveries by Briggs and coworkers and by Hager et al. (1993) therefore point to the important question of the possible relation between blue light-mediated phosphorylation changes and plasma membrane oxidoreductase activities. Moreover, blue light induces an electron transfer reaction between a plasma membrane flavoprotein and a *b*-type cytochrome (see the discussion on LIAC activity below).

An early discovered blue light-induced response that possibly indicated the primary action of the photoreceptor is related to membrane electron transport activities. Actinic blue light irradiation causes the reduction of a *b*-type cytochrome in fungal mycelia and membrane fractions (Asard and Caubergs 1991, Short and Briggs 1994). The light-induced absorbance change (LIAC) of this cytochrome was first detected in partially anaerobic mycelium from *Dictyostelium*, *Phycomyces* and *Neurospora* (Poff et al. 1973, Poff and Butler 1974, Munoz and Butler 1975) and was later demonstrated in higher plant subcellular fractions (Brain et al. 1977). The initially described absorbance changes also revealed the simultaneous reduction of a flavin molecule upon blue light irradiation. Subsequent action spectroscopy and studies using photochemical quenchers confirmed the action of a flavin chromophore in this reaction. Flavin-mediated LIAC activity has been further demonstrated in membrane fractions from *Neurospora* (Borgeson and Bowman 1985) and in developing cells from *Acetabularia* (Caubergs et al. 1984), supporting the ubiquitous nature of this photoreceptor system.

The most simple explanation of the photochemical events in the LIAC processes is probably a linear electron transport from a photoreduced flavin molecule to the *b*-type cytochrome. Although the mechanism of electron transfer between flavin and cytochrome has not yet been elucidated, it becomes increasingly clear that oxygen radicals may play an important role in this process. Studies on LIAC activity have resulted in a detailed analysis of the electrochemical properties of the major cytochrome *b* of plant plasma membranes (Asard et al. 1989, 1994) and the characterization of the involvement of ascorbic acid as a trans-plasma membrane electron donor to this system (Asard et al. 1992, Horemans et al. 1994). The molecules involved in LIAC activity are often referred to as the flavin-cytochrome *b*-system. It should be stressed that there is no direct evidence available today linking the LIAC phenomenon to physiological blue light photoperception. The physiological relevance of this system is therefore still a matter of debate (Widell 1987, Schmidt 1987).

The blue light-mediated activation of GTPase activity detected at the plasma membrane of pea apical buds (Warpeha et al. 1991) points to still another possible primary reaction of the blue light photoreceptor. Within 30-60 s after a short blue light pulse, a transient increase of GTPase activity was observed. Classical non-hydrolyzable inhibitors (GTP-g-S) completely abolished the reaction and red light was essentially ineffective in activating GTPase activity. It was concluded that a trimeric G-protein with a 40 kDa a-subunit was involved in the reaction. Moreover Warpeha et al. (1992) presented evidence that this reaction was mediated by a plasma membrane flavoprotein (see above). In any case, this blue light effect strongly indicates the possible transduction of blue light signals through typical hetero-trimeric G-proteins.

The plasma membrane of higher plants is also known to contain the elements of inositol phosphate metabolism (e.g. IP_3, diacylglycerol, see also Brody et al. this volume) including protein kinase C (for review see Drobak 1992). The inositol metabolism was suggested to be implicated in the signal transduction of *Samanea* pulvini (Morse et al. 1987). In addition a role for diacylglycerol and lipid activated protein kinase C in stomatal opening have been demonstrated by Lee and Assmann (1991). However, the possible coupling of blue light perception to altered inositol phosphate metabolism awaits further experimentation.

In summary, several second messenger systems may apparently be involved in the transduction of blue light signals. This situation likely reflects the heterogeneity of the blue light-mediated physiological responses.

7.4.3 Stomatal movements controlled by blue light

Stomatal opening is affected by blue light irradiation primarily through changes in K^+-uptake of the guard cells. K^+ transport similarly induces turgor changes in pulvini of *Samanea* and *Albizzia* causing rhythmic leaflet movements (Satter and Galston 1981). The transport of cations is apparently driven by H^+ extrusion from the motor cells originating from altered plasma membrane ATPase activity. Moreover the inhibition of the rhythmic leaflet movement by vanadate supports the involvement of the plasma membrane ATPase in this response (Saxe and Satter 1979).

Redox reactions at the plasma membrane of higher plants may however also mediate H^+ efflux in several systems (for review see Rubinstein and Luster 1993). A direct modulation by blue light of a guard cell plasma membrane oxidoreductase activity has even suggested (Raghavendra 1990). Blue light-sensitive redox activity at the plasma membrane of guard cell protoplasts was also proposed by Gautier et al. (1992) and these authors present a model indicating redox-mediated H^+-efflux. Thus it appears that the controversy on the involvement of the plasma membrane ATPase versus the plasma membrane reductases in H^+ efflux will not be easily settled. In an attractive hypothesis presented in detail by Marré et al. (1988), the oxidoreductases are responsible only for NAD(P)H-driven transmembrane electron transport with a concomitant membrane depolarization and acidification of the cytoplasm. These latter changes are likely to activate the plasma membrane ATPase resulting in increased H^+-efflux in a number of responses. Being the most prominent enzyme involved in the energization of many plasma membrane transport reactions, the plasma membrane ATPase is indeed expected to be sensitive to a number of environmental conditions including light (Roldan et al. 1991, see also Homblé this volume).

It should be stressed at this point that the blue light-mediated redox reactions mentioned here are by no means proven to involve the flavin-cytochrome *b* complex responsible for the plasma membrane LIAC activity. Initial attempts to correlate light-mediated cytochrome reduction and NAD(P)H-dependent redox activities have proven unsuccessful so far (Asard et al. 1987).

7.5 Is there a role for free flavins in blue light perception?

A hypothesis on the mechanism of action of blue light photoreception that has not usually be considered but deserves attention, specially as it is possibly related to CR synchronization, might involve a function for free cellular flavins. Soluble cytoplasmic flavin molecules reversibly associate with a specific binding protein that could potentially be located in a subcellular membrane (Hertel 1980, Ninnemann 1983). As such, the flavin-protein complex could function as a photoreceptor entity. High concentrations of free flavins, e.g. riboflavin, have indeed been found in certain cells, including the highly blue light-sensitive sporangiophore from *Phycomyces* (Hohl et al. 1992). There are two series of relevant experiments supporting this hypothesis.

First, a riboflavin-binding protein has been identified in several species. Membrane fractions from *Phycomyces* (Dohrmann 1983), *Zea, Cucurbita* (Hertel et al. 1980), *Phaseolus* (Asard, unpublished) and *Petroselinum* (Ensminger and Schäfer 1992) show a riboflavin-binding activity with an apparent affinity between 2 and 5 mM. The number of binding sites matches a theoretical minimal concentration calculated to generate a blue light photoresponse. It is unlikely that this activity merely reflects the presence of apoproteins devoid of their natural chromophores since FMN and FAD were significantly less effective in all cases. A riboflavin-binding site with similar kinetic properties was also demonstrated in highly purified plasma membranes from etiolated zucchini (*Cucurbita pepo*) and bean (*Phaseolus vulgaris*) hypocotyls pointing to its subcellular location (Caubergs unpublished). Additional riboflavin-binding sites in higher plants were found in a yet unidentified membrane fraction. Recent experiments also demonstrate the presence of an abundant riboflavin-binding protein in the photosensitive paraflagellar body of *Euglena* (Nebenführ et al. 1991). Although the exact location of this protein is still under investigation, it has been purified and detailed information on its structure may soon be available. The physiological function of this flavin-binding activity is still a puzzling question. Its abundance and widespread occurrence support a potential role in photoperception of blue light signals.

A second line of evidence supporting the free flavin-hypothesis is found in experiments on the CR of conidiation in *Neurospora crassa.* The conidiation response in the riboflavin-deficient double mutants *band rib1* and *band rib2* was sensitized by the addition of soluble flavins (Fritz et al. 1989, 1990). A comparison of several flavin derivatives clearly showed altered quantum efficiencies related to the photochemical properties of the molecules. The authors conclude that free flavin (i.e. riboflavin) controls the sensitivity of the circadian clock involved in synchronization of conidiation. Experiments using *Phycomyces* have suggested the possible role of free flavins in the sensitation of blue light signals (Otto et al. 1981). In addition a recent analysis by Hohl et al. (1992) indicated correlative changes between photogravitropic thresholds and endogenous FMN and FAD levels in several *Phycomyces* mutants. Therefore the role of free cellular flavins as a blue light photoreceptor in higher plants remains an attractive hypothesis that needs further investigation.

7.6 Conclusion

The understanding of the molecular mechanism by which plant cells perceive and transduce photomorphogenetically active signals is rapidly evolving. With respect to the perception of red light and of blue light new models have been recently proposed involving signal transduction mechanisms such as G-protein activation and protein phosphorylation. Although these transduction reactions are only implicated in specific responses such as phototropism and chloroplast development, they point to similarities with well known signal transduction systems from animal cells. Alternative approaches involving the analysis of photoreceptor mutants from higher plants and fungi are likely to complement this information.

Research on blue and red light photoperception has always considered the potential role of membrane transport processes in the early events of signal transduction. However, despite progress mentioned, little contribution has been made to the understanding of light-mediated changes in ion currents and membrane potentials. As evidenced by many examples throughout this book, also many rhythmic events in plants and fungi involve altered membrane transport properties. This by no means diminishes the importance of biochemical signal transduction pathways such as the activation of G-proteins and phosphodiesterase. Therefore, unravelling of the mechanism of induction of CR at the cellular level may provide essential information on the mechanism of action of non-photosynthetic photoreceptors.

Acknowledgements

The work in the laboratory of the authors is financially supported by the Belgian Fund for Scientific Research (NFWO). This support is greatfully acknowledged.

References

Ahmad M, Cashmore AR (1993) *HY4* gene of *A. thaliana* encodes a protein with characteristics of a blue-light photoreceptor. Nature 366:162-166

Antkowiak B, Mayer W-E, Engelmann W (1991) Oscillations of the membrane potential of pulvinar motor cells *in situ* in relation to leaflet movements of *Desmodium motorium*. J Exp Bot 42:901-910

Aparicio PJ, Quinones MA (1991) Blue light, a positive switch signal for nitrate and nitrite uptake by the green alga *Monoraphidium braunii*. Plant Physiol 95:374-378

Asard H, Caubergs RJ (1991) LIAC activity in higher plants. In: Lenci F, Ghetti F, Columbetti G, Häder D-P, Song P-S (eds) Biophysics of photoreceptors and photomovements in microorganisms. Plenum, New York, pp 181-189 (NATO ASI series, vol 211)

Asard H, Venken M, Caubergs RJ, Reijnders W, Oltmann FL, De Greef JA (1989) *b*-Type cytochromes in higher plant plasma membranes. Plant Physiol 90:1077-1083

Asard H, Horemans N, Briggs WR, Caubergs RJ (1995) Blue light perception by endogenous redox components of the plant plasma membrane. Photochem Photobiol (in press)

Asard H, Horemans N, Caubergs RJ (1992) Transmembrane electron transport in ascorbate-loaded plasma membrane vesicles from higher plants involve a *b*-type cytochrome. FEBS Lett 306:143-146

Asard H, Horemans N, Mertens J, Caubergs RJ (1994) The plant plasma membrane *b*-type cytochrome: a review. Bel J Bot 127:171-183

Asard H, Caubergs RJ, Renders D, De Greef JA (1987) Duroquinone-stimulated NADH oxidase and *b*-type cytochromes in the plasma membrane of cauliflower and mung beans. Plant Sci 53:109-119

Assman SM, Simoncini L, Schröder JI (1985) Blue light activates electrogenic ion pumping in guard cell protoplasts of *Vicia faba*. Nature 318:285-287

Berczi A, Asard H (1994) NAD(P)H-utilizing oxidoreductases of the plasma membrane; an overview of presently purified proteins. Protoplasma (in press).

Borgeson CE, Bowman BJ (1985) Blue light-reducible cytochromes in membrane fractions from *Neurospora crassa*. Plant Physiol 78:433-437

Bossen ME, Dassen HHA, Kendrick RE, Vredeberg WJ (1988) The role of calcium ions in phytochrome-controlled swelling of etiolated wheat (*Triticum aestivum* L.) protoplasts. Planta 174:94-100

Bowler C, Neuhaus G, Yamagata H, Chua N-H (1994) Cyclic GMP and calcium mediate phytochrome phototransduction. Cell 77:73-81

Brain RD, Freeberg JA, Weiss CV, Briggs WR (1977) Blue light-induced absorbance changes in membrane fractions from corn and *Neurospora*. Plant Physiol 59:948-952

Briggs WR, Iino M (1983) Blue-light-absorbing photoreceptors in plants. Phil Trans R Soc Lond B 303:347-359

Caubergs RJ, Vanden Driessche T, De Greef JA (1984) A light inducible cytochrome *b* reduction in the green alga *Acetabularia*. In: Senger H (ed) Blue light effects in biological systems. Springer-Verlag, Berlin Heidelberg New York. pp 173-176

Chory J (1994) Phytochrome signal transduction. Curr Biol 4:844-846

Comolli J, Taylor W, Hasting JW (1994) An inhibition of protein phosphorylation stops the circadian oscillator and blocks light-induced phase shifting in *Gonyaulax polyedra*. J Biol Rhythms 9:13-26

Dohrman U (1983) In)vitro riboflavin binding and endogenous flavins in *Phycomyces blakesleeanus*. Planta 159:357-365

Dring MJ (1988) Photocontrol of development in algae. Anny Rev Plant Physiol Plant Mol Biol 39:157-174
Drobak BK (1992) The plant phosphoinositide system. Biochem J 288:135-150
Edgerton MD, Jones AM (1992) Localization of protein-protein interactions between subunits of phytochrome. Plant Cell 4:161-171
Ensminger P, Schäfer E (1992) Blue and ultraviolet-B light photoreceptors in parsley cells. Photochem Photobiol 55:437-447
Feldman JF, Dunlap JC (1983) *Neurospora crassa*: a unique system for studying circadian rhythms. In: Smith KC (ed) Photochemical and photobiological reviews. Plenum Press, New York, pp 319-368
Fritz BJ, Kasai S, Matsui K (1990) Blue light photoperception in *Neurospora* circadian rhythm: evidence for involvement of the flavin triplet state. Photochem Photobiol 51:607-610
Fritz BJ, Kasai S, Matsui K (1989) Free cellular riboflavin is involved in phase shifting by light of the circadian clock in *Neurospora crassa*. Plant Cell Physiol 30:557-564
Fukshansky L, Richter T (1990) What do we know about the non-uniform perception of a phototropic stimulus in *Phycomyces*? Planta 182:107-112
Furuya M (1993) Phytochromes: their molecular species, gene families and functions. Annu Rev Plant Physiol Plant Mol Biol 44:617-645
Furuya M, Freer JH, Ellis A, Yamamoto KT (1981) Electrostatic binding of proteins and phytochrome to differently charged liposomes. Plant Cell Physiol 22:135-144
Furuya M, Wada M, Kadota A (1980) Regulation of cell growth and cell cycle by blue light in *Adiantum* gametophytes. In: Senger H (ed) The blue light syndrome. Springer-Verlag, Berlin. pp 119-132
Gallagher S, Short T, Ray PM, Pratt LH, Briggs WR (1988) Light-mediated changes in two proteins found associated with plasma membrane fractions from pea stem sections. Proc Natl Acad Sci USA 85:8003-8007
Galland P (1990) Phototropism of the *Phycomyces* sporangiophore: a comparison with higher plants. Photochem Photobiol 52:233-248
Galland P, Lipson ED (1987) Blue light reception in *Phycomyces* phototropism: evidence for two photosystems operating in low- and high-intensity ranges. Proc Natl Acad Sci USA 84:104-408
Galland P, Lipson ED (1984) Photophysiology of *Phycomyces blakesleeanus*. Photochem Photobiol 40:795-800
Galland P, Senger H (1988) The role of pterins in the photoperception and metabolism of plants. Photochem Photobiol 48:811-820
Gautier H, Vavasseur A, Lasceve G, Boudet AM (1992) Redox processes in the blue light response of guard cell protoplasts of *Commelina commnis* L. Plant Physiol 98:34-38
Goldsmith MHM, Caubergs RJ, Briggs WR (1980) Light-inducible cytochrome reduction in membrane preparations from corn coleoptiles. I. Stabilization and spectral characterization of the reaction. Plant Physiol 66:1067-1073
Hager A, Brich M (1993) Blue light-induced phosphorylation of a plasma membrane protein from phototropically sensitive tips of maize coleoptiles. Planta 189:567-576
Hager A, Brich M, Bazlen I (1993) Redox dependence of the blue light-induced phosphorylation of a 100-kDa protein in isolated plasma membranes from tips of coleoptiles. Planta 190:120-126
Halle CC, Roux SJ (1980) Photoreversible calcium fluxes induced by phytochrome in oat coleoptile cells. Plant Physiol 65:658-662
Hanstein C, Grolig F, Wagner G (1992) Immunolocalization of cytosolic phytochrome in the green alga *Mougeotia*. Bot Acta 105:55-62
Haupt W (1982) Light-mediated movement of chloroplasts. Annu Rev. Plant Physiol 33:205-233

Hertel R, Jesaitis A, Dohrmann U, Briggs WR (1980) In vitro binding of riboflavin to subcellular particles from maize coleoptiles and *Cucurbita* hypocotyls. Planta 147:312-319

Hertel R (1980) Phototropism of lower plants. In: Lenci F, Colombetti G, Song P-S (eds) Photoperception and sensory transduction in aneural organisms. Plenum Press, New York, pp 89-95 (NATO ASI series, vol 89)

Hohl N, Galland P, Senger H (1992) Altered flavin patterns in photobehavioral mutants of *Phycomyces blakesleeanus*. Photochem Photobiol 55:247-255

Horemans N, Asard H, Caubergs RJ (1994) The role of ascorbate free radical as an electron acceptor to cytochrome *b*-mediated trans-plasma membrane electron transport in higher plants. Plant Physiol 104:1455-1458

Iino M, Ogawa T, Zeiger E (1985) Kinetic properties of the blue-light response of stomata. Proc Natl Acad Sci USA 82:8019-8023

Kaufman LS (1993) Transduction of blue light signals. Plant Physiol 102:333-337

Kaufman LS (1994) GTP-binding signalling proteins in higher plants. J Photochem Photobiol B: Biol 22:3-7

Kim HY, Coté GG, Crain RC (1992) Effects of light on the membrane potential of protoplasts from *Samanea saman* pulvini. Plant Physiol 99:1532-1539

Kondo T, Kubota M, Aono Y, Watanabe M (1988) A computerized video system to automatically analyze movements of individual cells and its application to the study of circadian rhythms in phototaxis and motility of *Chlamydomonas reinhardtii*. Protoplasma Suppl 1:185-192

Konjevic R, Khurana JP, Poff KL (1992) Analysis of multiple photoreceptors pigments for phototropism in a mutant of *Arabidopsis thaliana*. Photochem Photobiol 55:789-792

Konjevic R, Steinitz B, Poff KL (1989) Dependence of the phototropic response of *Arabidopsis thaliana* on fluence rate and wavelength. Proc Natl Acad Sci USA 86:9876-9880

Lamparter T, Lutterbüse P, Schneider-Poetsch HAW, Hertel R (1992) A study of membrane-associated phytochrome: hydrophobicity test and native size determination. Photochem Photobiol 56:697-707

Lee Y, Assmann SM (1991) Diacylglycerols induce both ion pumping in patch-clamped guard cell protoplasts and opening of intact stomata. Proc Natl Acad Sci USA 88:2127-2131

Leong T-Y, Vierstra RD, Briggs WR (1981) A blue light-sensitive cytochrome-flavin complex from corn coleoptiles. Further characterization. Photochem Photobiol 34:697-703

Lopez-Figueroa F, Rüdiger W (1991) Stimulation of nitrate uptake and reduction by red and blue light and reversion by far-red light in the green alga *Ulva rigida*. J Phycol 27:389-394

Lumsden PJ (1991) Circadian rhythms and phytochrome. Annu Rev Plant Physiol Plant Mol Biol 42:351-371

Lumsden PJ, Saji H, Furuya M (1987) Action spectrum confirm two separate actions of phytochrome in the induction of flowering in *Lemna paucicostata* 411. Plant Cell Physiol 28:1237-1242

Manabe K, Furuya M (1975) Distribution and non-photochemical transformation of phytochrome in subcellular fractions from *Pisum* epicotyls. Plant Physiol 56:772-775

Marré MT, Moroni A, Albergoni FG, Marré E (1988) Plasmalemma redox activity and H^+ extrusion. I. Activation of the H^+-pump by ferricyanide induced potential depolarization and cytoplasm acidification. Plant Physiol 87:25-29

Morse MJ, Crain RC, Satter RL (1987) Light stimulated phosphatidylinositol turnover in *Samanea saman* pulvini. Proc Natl Acad Sci USA 84:7075-7078

Munoz V, Butler WL (1975) Photoreceptor pigment for blue light in *Neurospora crassa*. Plant Physiol 55:421-426

Nebenführ A, Schäfer A, Galland P, Senger H, Hertel R (1991) Riboflavin-binding sites associated with flagella of *Euglena*: a candidate for blue-light photoreceptor? Planta 185:65-71

Neuhaus G, Bowler C, Kern R, Chua N-H (1993) Calcium/calmodulin-dependent and -independent phytochrome signal transduction pathways. Cell 73:937-952

Ninneman H (1983) Reversible absorbance changes and modulation of biological activities by blue light. In: Montagnoli G, Erlanger BF (eds) Molecular models of photoresponsiveness. Plenum Press, New York. pp 133 (NATO ASI series, vol 68)

Otto MK, Jayaram M, Hamilton RM, Delbrück M (1981) Replacement of riboflavin by an analogue in the blue-light photoreceptor of *Phycomyces*. Proc Natl Acad Sci USA 78:266-269

Paietta J, Sargent ML (1983) Modification of blue light photoresponses by riboflavin analogs in *Neurospora crassa*. Plant Physiol 72:764-770

Palmer JM, Short TW, Briggs WR (1993) Correlation of blue light-induced phosphorylation to phototropism in *Zea mays* L. Plant Physiol 102:1219-1225

Poff KL, Butler WL, Loomis WF (1973) Light induced absorbance changes associated with phototaxis in *Dictyostelium discoideum*. Proc Natl Acad Sci USA 57:813-816

Poff KL, Butler WL (1974) Absorbance changes induced by blue light in *Phycomyces blakesleeanus* and *Dictyostelium discoideum*. Nature 248:799-801

Pratt LH (1979) Phytochrome: function and properties. Photochem Photobiol Rev 4:59-124

Pratt LH (1995) Phytochromes: differential properties, expression patterns and molecular evolution. Photochem Photobiol 61:10-21

Quail PH (1991) Phytochrome: a light-activated molecular switch that regulates plant gene expression. Annu Rev Genet 25:389-409

Quinones MA, Zeiger E (1994) A putative role of the xanthophyll, zeaxanthin, in blue light photoperception of corn coleoptiles. Science 264:558-561

Racusen RH, Cooke TJ (1982) Electrical changes in the apical cell of the fern gametophyte during irradiation with photomorphogenetically active light. Plant Physiol 70:331-334

Raghavendra AS (1990) Blue light effects on stomata are mediated by the guard cell plasma membrane redox system distinct from the proton translocating ATPase. Plant Cell Environm 13:105110

Ramirez JM, Gallego GG, Serrano R (1984) Electron transfer constituents in plasma membrane fractions from *Avena sativa* and *Saccharomyces cerevisea*. Plant Sci 34:103-110

Ranjeva R, Boudet AM (1987) Phosphorylation of proteins in plants: regulatory effects and potential involvement in stimulus/response coupling. Annu Rev Plant Physiol 38:73-93

Raven JA (1983) Do plant photoreceptors act at the membrane level? Phil Trans R Soc Lond B. 303:403-417

Reymond P, Short TW, Briggs WR (1992) Blue light activates a specific protein kinase in higher plants. Plant Physiol 100:655-661

Roldan M, Donaire JP, Pardo JM, Serrano R (1991) Regulation of root plasma membrane H^+-ATPase in sunflower seedlings. Plant Sci 79:163-172

Rubinstein B, Luster DG (1993) Plasma membrane redox activity: components and role in plant processes. Annu Rev Plant Physiol Plant Mol Biol 44:131-155

Russo VEA (1988) Blue light induces circadian rhythms in the *bd* mutant of *Neurospora*: double mutants *bd, wc-1* and *bd, wc-2* are blind. J Photochem Photobiol B: Biol 2:59-65

Satter RL, Galston AW (1981) Mechanism of control of leaf movements. Anny Rev Plant Physiol 32:82-110
Satter RL, Guggino S, Lonegran TA, Galston AW (1981) The effects of blue and far red light on rhythmic leaflet movements in *Samanea* and *Albizzia*. Plant Physiol 67:965-968
Saxe H, Satter RL (1979) Effect of vanadate on rhythmic leaflet movement in *Albizzia julibrissin*. Plant Physiol 64:905-907
Schmid R (1987) Blue light effects in endogenous rhythms. In: Senger H (ed) Blue light responses: phenomena and occurrence in plants and microorganisms. CRC Press, Boca Raton, Vol 1, pp 87-99
Schmidt W (1987) Primary reactions and optical spectroscopy of blue light photoreceptors. In: Senger H (ed) Blue light responses: phenomena and occurrence in plants and microorganisms. CRC Press, Boca Raton, Vol 2, pp 19-36
Schröder-Lorenz A, Rensing L (1987) Circadian changes in protein synthesis rate and protein phosphorylation in cell-free extracts of *Gonyaulax polyedra*. Planta 170:7-13
Senger H (ed) (1984) Blue light effects in biological systems. Springer-Verlag, Berlin Heidelberg New York.
Senger H (ed) (1987) Blue light responses: phenomena and occurrence in plants and microorganisms. CRC Press, Boca Raton
Sharrock RA, Quail P (1989) Novel phytochrome sequences in *Arabidopsis thaliana*: structure, evolution and differential expresion of a plant regulatory photoreceptor family. Genes Dev 3:1745-1757
Shimazaki K, Iino M, Zeiger E (1986) Blue-light-dependent proton extrusion by guard-cell protoplasts of *Vicia faba*. Nature 319:324-326
Short TW, Briggs WR (1990) Characterization of a rapid, blue light-mediated change in detectable phosphorylation of a plasma membrane protein from etiolated pea (*Pisum sativum* L.). Plant Physiol 92:179-185
Short TW, Briggs WR (1994) The transduction of blue light signals in higher plants. Ann Rev Plant Physiol Plant Mol Biol 45:143-171
Sineshchekov V, Lamparter T, Hartmann E (1994) Evidence for the existence of membrane-associated phytochrome in the cell. Photochem Photobiol 60:516-520
Song P-S (1984) Photophysical aspects of blue light receptors: the old question (flavins versus carotenoids) re-examined. In: Senger H (ed) (1984) Blue light effects in biological systems. Springer-Verlag, Berlin Heidelberg New York, pp 75-80
Steinhardt AR, Popescu T, Fukshansky L (1989) Is the dichroic photoreceptor for *Phycomyces* phototropism located at the plasma membrane or at the tonoplast? Photochem Photobiol 49:79-87
Takagi S, Yamamoto KT, Furuya M, Nagai R (1990) Cooperative regulation of cytoplasmic streaming and Ca^{2+} fluxes by Pfr and photosynthesis in *Vallisneria* mesophyll cells. Plant Physiol 94:1702-1708
Takagi S, Nagai R (1988) Light-affected Ca^{2+} fluxes in protoplasts from *Vallisneria* mesophyll cells. Plant Physiol 88:228-232
Terryn N, Van Montagu M, Inzé D (1993) GTP-binding proteins in plants. Plant Mol Biol 22:143-152
Thomas B (1991) Phytochrome and photoperiodic induction. Physiol Plant 81:571-577
Vanden Driessche Th (1994) Circadian rhythms in three unicellular organisms. The peculiarities of the organisms, the evidence brought on rhythms and their specific practical problems. Outline of recent hypotheses. Biol Rhythm Res 25:345-386
Vierstra RD (1993) Illuminating phytochrome functions. There is light at the end of the tunnel. Plant Physiol 103:679-684
Warpeha KMF, Hamm HE, Rasenick MM, Kaufman LS (1991) A blue-light-activated GTP-binding protein in the plasma membrane of etiolated peas. Proc Natl Acad Sci USA 88:8925-8929

Warpeha KMF, Kaufman LS, Briggs WR (1992) A flavoprotein may mediate the blue light-activated binding of guanosine 5'-triphosphate to isolated plasma membranes of *Pisum sativum* L. Photochem Photobiol 55:595-603
Widell S (1987) Membrane-bound blue light receptors. Possible connection to blue light photomorphogenesis. In: Senger H (ed) Blue light responses: phenomena and occurrence in plants and microorganisms. CRC Press, Boca Raton, Vol 2, pp 89-98

8 Temporal Variations in Tissular and Plasma Protein Binding of Drugs With Special Reference to their Erythrocyte Penetration as Mechanisms Involved in Chronopharmacological Effects

Bernard Bruguerolle.
Université d'Aix-Marseille II, Laboratoire de Pharmacologie Médicale et Clinique, Faculté de Médecine de Marseille.

Summary

Temporal variations in the mode of action of the drug (i.e.chronopharmaco dynamics) as well as temporal variations in its pharmacokinetics (i.e. chronokinetics) may explain temporal variations in drug effects .

Studies on the penetration of drugs into erythrocytes as well as studies on their plasma protein and tissular binding document temporal variations of membrane permeability to drugs.

In animals as well as in man several studies have described circadian variations in drug protein binding for drugs such as antiepileptics, antiarrhythmic drugs, local anaesthetics...

Concerning local anaesthetics for instance we have documented circadian variations in plasma protein binding which may explain the temporal variations in drug kinetics previously reported.

By using the erythrocytic passage as a tool for the study of temporal variations in drug permeability, we have also documented circadian variations in erythrocyte penetration of local anaesthetics.

Finally, circadian variations in tissular drug penetration may also provide an indirect reflect of temporal variations in tissular drug permeability.

8.1. Introduction

The effect and fate of a drug in an organism which comprises multiple biological rhythms (i.e. its temporal structure) depend on its time of administration; chronopharmacology studies temporal variations of the activity, toxicity and kinetics of drugs according to their time of administration and possible alterations of the temporal structure of the organism after drug treatment (Bruguerolle 1989 a, b, Bruguerolle, 1988, Reinberg, 1992).

Pharmacology is concerned with the effectiveness of drugs; its mode of action i.e. pharmacodynamics and its pharmacokinetics (study of the fate of drugs in the organism). These two aspects interfere since the response of the receptor depends on the concentration of the drug at the receptor site (involving pharmacokinetics processes) as well as on the fate of the drug (i.e. its absorption, its distribution, its metabolism and its elimination) .

To be effective a drug must be in appropriate concentrations at its sites of action; the concentrations reached depend on the rate of its absorption, distribution, binding or localization in tissues, biotransformations and excretion. All these steps involve its passage across cell membranes : to permeate a cell, a drug must cross the cellular plasma membrane either by a passive process or by a mechanism involving active participation of membrane components. The plasma membrane is involved in intra and extracellular exchanges and represents a preferential site of action for numerous drugs.

Thus, as far as chronopharmacology is concerned, both temporal variations in the mechanisms by which a drug crosses the membranes and its physico-chemical properties influence this transfer and may be involved in chronodynamics and chronokinetics.

The study of distribution of drugs and especially tissular and plasma protein binding studies may provide indirect evidence for these temporal variations .

The present chapter aims to explain by some experimental and clinical chronopharmacological data from our group (particularly local anaesthetics) that temporal variations in tissular, erythrocyte and plasma protein binding result at least partially from chronopharmacological effects [3].

8.2. Temporal variations in plasma protein binding of drugs

Drugs are transported from their site of administration to their receptors by plasma proteins such as serum albumin, alpha 1 acid glycoprotein, globulins and lipoproteins. The unbound drug represents the active fraction since it is the only

---

[3]Definitions for pharmacokinetic parameters: Cmax.: maximum blood or plasma concentration, T max.: time to reach Cmax., $T_{1/2}\ \beta$: half-life of the drug elimination phase.

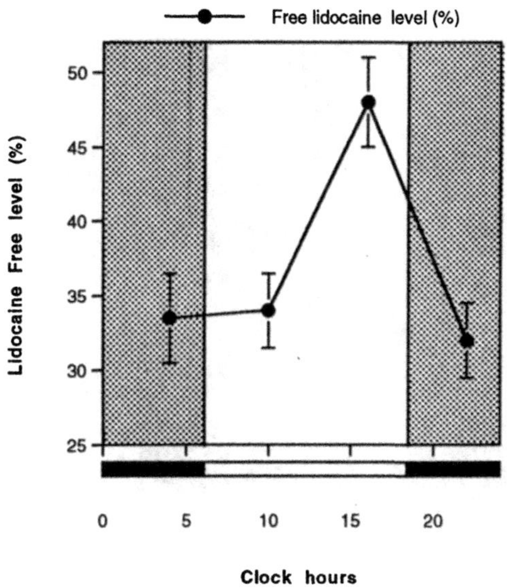

Fig. 8.1. Temporal variations in lidocaine free plasma levels in rats (Redrawn from Bruguerolle et al., 1983)

fraction that can diffuse through membranes and tissues; thus, protein binding of drugs has important pharmacological implications.

Drug protein binding may be influenced by temperature, pH, physico-chemical properties, plasma concentration of the involved protein... each of these factors being theoretically subjected to temporal variations.

In order to try to explain some of our chronokinetic findings, we investigated temporal variations in plasma protein binding of various drugs: thus, we reported circadian variations for the free plasma drug levels (and accordingly for plasma protein binding) of acid drugs (carbamazepine) or basic drugs (bupivacaine, etidocaine, lidocaine, mepivacaine, disopyramide).

1- The very first study concerned the chronokinetics of carbamazepine in rats (Bruguerolle et al., 1981). We demonstrated that the highest values for protein binding of this drug were observed at 04.00 h during the dark phase (LD:12/12 with L from 06.00 h to 18.00 h).

2- In a second study (Bruguerolle et al., 1983), the chronokinetics of lidocaine were investigated at four fixed different time points in a 24 hour period , in rats after a single intramuscular injection of this drug (LD:12/12 with L from 06.00 h to 18.00 h); circadian variations of lidocaine kinetics were demonstrated: when lidocaine was given at 16.00 h its elimination half-life was shorter and the peak (C_{max}) was higher than when given at 10.00 h, 22.00 h or 04.00 h. These findings may explain, in part at least, that susceptibility of mice to lidocaine is time dependent, as reported by Lutsch and Morris (1967) and our group (Bruguerolle and Emperaire, 1993). Among the possible temporal variations of distribution, metabolism and elimination processes that may be responsible for such varia-

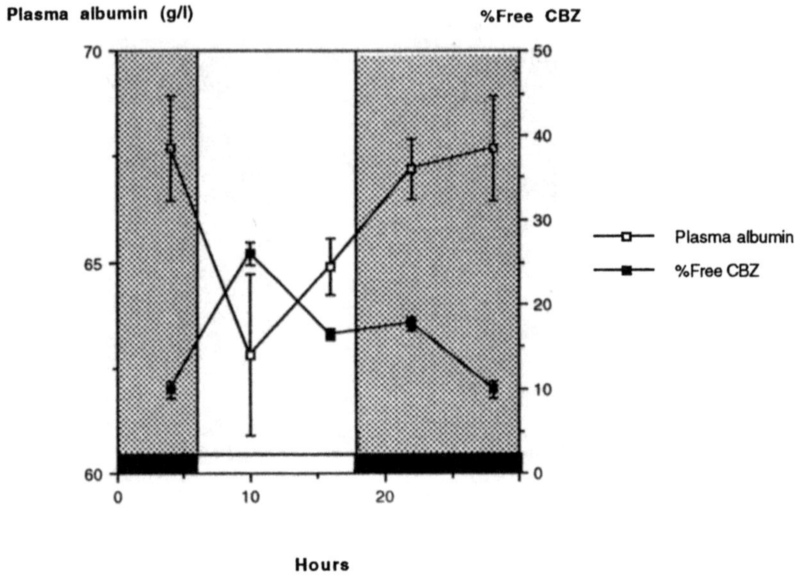

Fig. 8.2. Temporal variations in carbamazepine free plasma levels and albumin concentrations in rats (redrawn fromBruguerolle et al.,1981)

tions, circadian variations of membrane permeability may be involved as well and were thus investigated. The highest free plasma levels of lidocaine were found at 16.00 h as shown in figure 1 ; on the contrary, the lowest values (i.e. maximum values for protein binding) occured during the dark phase.

3- Similar chronokinetics of bupivacaine were reported in mice (Bruguerolle and Prat, 1987): after a single 20 mg/kg intraperitoneal injection of this drug at 10.00 h, 16.00 h, 22.00 h or 04.00 h pharmacokinetic parameters were found to depend on the hour of administration; highest Cmax, Cmax/Tmax ratio and longest $T_{1/2}$ ß occured at 22.00 h (Bruguerolle and Prat,1987) as well as the highest free fraction which was observed at 22.00 h.

The same results were obtained after mepivacaine injection in mice (Cmax/Tmax ratio and longest $T_{1/2}$ ß observed at 22.00h)(Bruguerolle and Prat, 1988 a) and here as well the maximum free fraction was observed at 10.00 h.

Finally, concerning etidocaine, the highest total plasma Cmax occured at 04.00 h and the highest free fraction was observed at 04.00 h (Bruguerolle and Prat, 1990 a).

Concerning all these local anaesthetics, our data also showed that the highest toxicity of the drug was observed at the highest plasma levels obtained (22.00 h or 04.00 h) (Bruguerolle and Prat,1988 c, Prat and Bruguerolle, 1988).

It must be pointed out that temporal changes in plasma protein binding are usually reported to be dependent on those in the amount of plasma proteins i.e. albumin (the protein specifically responsible for plasma binding of carbamazepine) and alpha 1 acid glycoprotein (responsible for plasma binding of local anaesthetics). Plasma total proteins and albumin were shown to vary along the

24-h scale with a maximum during the dark phase of the light/dark cycle (Valli et al., 1979): thus, as far as carbamazepine is concerned, the temporal variations of albumin were demonstrated to be inversely correlated to those of the carbamazepine free fraction as shown in figure 2.

Although circadian variations in alpha 1 acid glycoprotein have not yet been established in rodents, it may be assumed that such a protein varies with a maximum occuring during the dark phase as observed for most of the other plasma proteins in rodents (Valli et al., 1979). Once more, temporal variations of the amount of the main proteins responsible for plasma binding of local anaesthetics should explain variations in protein binding of these drugs.

Until now few studies have been devoted to temporal variations in plasma protein binding of drugs in humans (Bruguerolle and Lemmer, 1993, Lemmer and Bruguerolle, 1994). The highest plasma levels of free phenytoin or valproic acid were observed in man between 02.00 h and 06.00 h (Patel et al.,1982, Lockard et al.,1985). the lowest levels of free diazepam (Naranjo et al.,1980) and carbamazepine (Riva et al.,1984) were reported to occur in the morning. Also, cisplatin binding to plasma proteins varied along the 24-hour scale (maximum in the afternoon and minimum in the morning) as described by Hecquet et al. (1984). Bauer et al. (1985) reported on circadian variations of valproic acid protein binding in young and elderly subjects: the clearance of the unbound drug was about 15% higher during the evening.

It is well known that different plasma proteins such as albumin, alpha 1 acid glycoprotein, are concerned with protein binding of drugs. Thus, temporal variations of various plasma proteins may be involved in such variations as reported by Reinberg et al. (1977) or Bruguerolle et al. (1989) . The physiological circadian time structure of some plasma proteins in man and its alterations by age or different disease states such as cancer (Bruguerolle et al.,1989) or inflammatory diseases (Focan et al., 1988) were studied (figure 3).

As noted above, temporal changes are usually reported to be dependent on those of the concentration of plasma proteins but not on possible temporal changes in the affinity of the proteins concerned: to our knowledge such studies were not performed until now; thus future studies should document temporal changes in the protein binding affinity characteristics of drugs.

Regarding the possible explanations of chronopharmacological effects or of chronokinetic mechanisms, such variations may be of particular importance.

More recently we documented circadian variations in plasma protein binding of three local anaesthetics (bupivacaine, etidocaine and mepivacaine) in mice and tried to correlate these variations with temporal variations in toxicity and kinetics of these drugs (Bruguerolle and Prat, 1992).A significant circadian variation in protein binding of the three local anaesthetics was demonstrated. As for the maximum concentration of total plasma level, the highest free plasma level occured at the same time that the hour of maximum acute toxicity for bupivacaine (22.00 h) and etidocaine (04.00 h) but not for mepivacaine: the temporal variation of free plasma levels may explain, at least partially for bupivacaine and etidocaine, the temporal changes of local anaesthetic induced mortality.

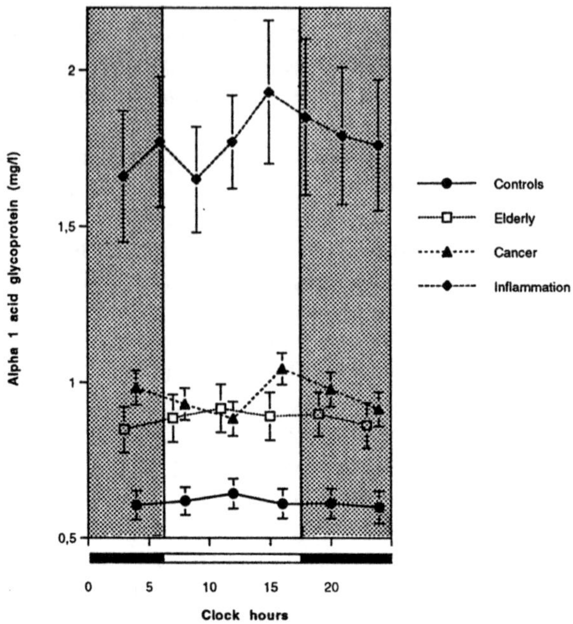

Fig. 8.3. Circadian time structure of alpha one acid glycoprotein in healthy young men, elderly, cancerous patients and patients with inflammatory diseases (redrawn from Bruguerolle et al., 1989)

The clinical applications of such temporal variations of plasma protein drug binding can be questionned. It is generally admitted that clinically significant consequences of changes in drug binding are only observed for drugs which are highly bound (more than 80 %). Thus, it can be assumed that temporal variations in plasma drug binding may have clinical implications only for highly bound drugs characterized by a small apparent volume of distribution.

8.3. Temporal variations in tissular drug binding

Tissue penetration regulates the effectiveness and toxicity of many drugs. Although most information about distribution of drugs can be derived from measurement of blood drug concentrations, the study of distribution of a drug also requires its measurement in various tissues. Many factors including the delivery of the drug to the tissues, the passage through tissue membranes and the binding of the drug to plasma proteins and tissue components are involved in the distribution pattern of a drug.

Some experimental data have been reported on tissular penetration of drugs such as betablocking agents, antidepressants (Lemmer et Holle,1991).

Circadian variations in tissue binding of three local anaesthetics were investigated by our group: time dependency in tissular (heart and brain) levels of bupivacaine, etidocaine and mepivacaine were documented in mice; taking into account the neurological and cardiovascular toxicity of these drugs, this point is of a particular importance. Thus, we have demonstrated a circadian variation of the penetration of these three local anaesthetics in heart and brain tissues (Prat and Bruguerolle, 1990 a,b, Bruguerolle and Prat, 1992): the maximum passage was observed at 10.00 h in cardiac tissues for the three drugs and at 10.00 h, 16.00 h and 22.00 h in brain tissues for bupivacaine, etidocaine and mepivacaine, respectively .

Nevertheless, we did not demonstrate a temporal relationship between the respective free and tissular levels; thus the temporal variation of free plasma, brain and heart levels may participate to chronokinetics but does not seem to explain the temporal changes of local anaesthetic induced mortality.

8.4. Penetration of drugs into erythrocytes as a tool for the study of temporal variation in the passage of drugs through membranes

Erythrocytes are usually chosen as a model for membrane permeability studies("in vitro" and "ex vivo"). Drugs may be transported by red blood cells: this binding may depend on the plasma protein binding of drugs and on their physicochemical properties.

Temporal changes in membrane permeability of drugs have been proposed as an explanation of both chronopharmacodynamics and chronopharmacokinetics.

Thus the time dependency of the penetration of drugs into red blood cells may provide a strong argument for the existence of temporal variations in the transport of drugs across biological membranes.

To our knowledge, time dependency of drug binding to erythrocytes has not yet been reported with the exception of that of local anaesthetics (lidocaine, bupivacaine, etidocaine and mepivacaine) in rodents and of indomethacin in human elderly volunteers. The possible temporal changes of the penetration of several drugs through the erythrocyte membrane were investigated by the determination of their respective erythrocyte concentrations, according to the hour of administration.

We reported in rats the chronokinetics of lidocaine as well as the circadian variations of its passage into red blood cells (Bruguerolle et al.,1983). To assess the circadian time dependent passage into the erythrocytes, the ratio of drug concentrations in erythrocytes to plasma drug concentrations was calculated; when the drug was given at 22.00 h the ratio was 0.74 ± 0.04 versus 0.48 ± 0.02 when given at 10.00 h (Bruguerolle et al.,1983).

Temporal variations in the passage of bupivacaine, etidocaine and mepivacaine into red blood cells were also investigated in mice (Bruguerolle and Prat, 1988 b,

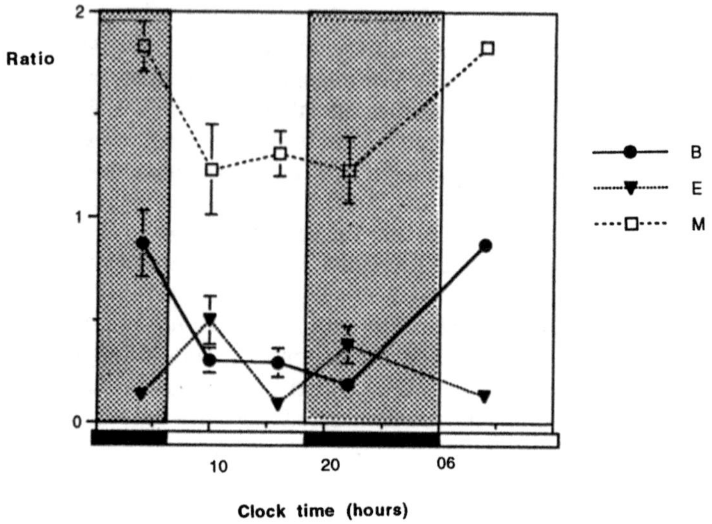

Fig. 8.4. Temporal variations in penetration of three local anaesthetic agents (bupivacaine, etidocaine and mepivacaine) into erythrocytes (ratio of erythrocytic vs plasma concentrations) in mice (redrawn from Bruguerolle and Prat., 1992).

Bruguerolle and Prat , 1990 b). A circadian variation in the passage of the three anaesthetic agents was demonstrated with a maximum occuring at 04.00 h for bupivacaine and at 10.00 h for etidocaine and mepivacaine (figure 4).

It has also been reported that, in rats, time dependency of drug binding to erythrocytes varies according to another period than the 24-hours scale i.e. estrous cycle; we demonstrated that the kinetics of theophylline depend on the phase of the estrous cycle (Bruguerolle, 1987 a) and that moreover the passage of this drug into red blood cells was dependent on the phase of the estrous cycle (Bruguerolle, 1987 b).

8.5. Chronopharmacological mechanisms at receptor sites

To be effective, most drugs must be in appropriate concentrations at their sites of action; a chronopharmacological approach of the action of drugs at their receptor sites involves the study of temporal variations in the number and/or the affinity of their binding sites. The charactéristics of some studies, carried out in animals or in human subjects, of various drugs (opiates, benzodiazepines, insulin, imipramine, alpha or beta receptor blocking drugs, oestrogens...) for several periodicities (circadian, mensual,circannual) and at a central or a peripheral level, have been reviewed elsewhere (Bruguerolle, 1992).

All these studies clearly indicate that for a given receptor the number of sites avaiable or its affinity sites may vary according to time and are not constant throughout the day, the month or the year: the changes at the receptor level may explain, in part, some of the chronopharmacological effects.

Concerning the local anaesthetic agents previously mentionned, the mechanism of their action involves a block in nerve conduction by decreasing or preventing the large transient increase in the permeability of the membrane to sodium, produced by a slight depolarization of the membrane.

Thus, drug-induced changes, by local anaesthetics, in the membrane permeability of nerve cells to ions are the main mode of action of theses drugs.

Circadian changes in the effects of local anaesthetics may result from temporal changes in membrane properties as suggested by Reinberg and Reinberg (1977). Circadian changes in membrane permeability and circadian changes in potassium, sodium or calcium efflux from cells may explain, in part at least, temporal changes in local anaesthetic efficiency and kinetics.

Dorian et al. (1988) documented such variations in mice by using the erythrocyte as a model: possible temporal variations in erythrocyte membrane permeability to sodium and potassium were documented; plasma as well as erythrocyte sodium and potassium concentrations were found to vary along the 24-hour scale as well as the erythrocytic sodium/potassium ratio. A large increase of intracellular sodium was detected at 02.00 h. These data suggested a possible circadian variation in the ATPase activity of the red blood cell membrane .

8.6. Conclusions

In terms of toxicity, efficiency and/or kinetics, the temporal variations demonstrated for several anaesthetic agents may be explained, in part at least, by temporal changes in membrane permeability.

Indirect evidence for temporal variations in membrane permeability have been documented by the study of distribution processes of local anaesthetics and especially by plasma protein binding, tissular (heart and brain) concentrations and erythrocyte penetration.

Acknowledgments

Special thanks to Manon Gantenbein for assistance in the preparation of the manuscript

References

Bauer LA, Davis R, Wilensky A, Raisys V, Levy RH (1985) Valproic acid clearance: unbound fraction and diurnal variation in young and elderly adults. Clin Pharm Ther 37: 697-700

Bruguerolle B (1987 a) Modifications de la pharmacocinétique de la théophylline au cours du cycle oestral chez la ratte. Pathol Biol 35: 181-183

Bruguerolle B (1987 b) Erythrocyte concentrations of theophylline:influence of the oestrous cycle. Med Sci Research-Biochem 15: 263-264

Bruguerolle B (1989 a) Time dependence of general and localanaesthetic drugs.In: Lemmer B (ed) Chronopharmacology, cellular and biochemical interactions. Marcel Dekker Inc, New York p 581

Bruguerolle B.(1989 b).Temporal aspects in drug absorption and drug distribution. In: Lemmer B (ed) Chronopharmacology, cellular and biochemical interactions. Marcel Dekker Inc, New York p 3

Bruguerolle B (1992) Chronopharmacology. In: Touitou Y, Hauss E (eds), Biological rhythms in clinical and laboratory medicine, Springer-Verlag p 114

Bruguerolle B, Arnaud C, Levi F, Focan C, Touitou Y, Bouvenot G (1989) Physiopathological alterations of alpha 1 acid glycoprotein temporal variations: implications for chronopharmacology. In: Baumann P et al. (eds), Alpha 1 acid glycoprotein: genetics, biochemistry, physiological functions and pharmacology. Alan R. Liss,Inc.New York p 199

Bruguerolle B. and Emperaire N. (1993). Flumazenil and lidocaine-induced toxicity: is the inverse agonist type activity circadian-time dependent ? J Pharm Pharmacol 45:678-679

Bruguerolle B, Jadot G (1983) Influence of the hour of administration of lidocaine on its intraerythrocytic passage in the rat. Chronobiologia 10: 295-297

Bruguerolle B, Lemmer B (1993). Recent advances in chronopharmacokinetics-Methodological problems. Life Sci 52: 1809-1824

Bruguerolle B, Prat M (1987) Temporal changes in bupivacaine kinetics. J Pharm Pharmacol 39: 148-149

Bruguerolle B, Prat M (1988 a). Circadian phase dependent pharmacokinetics and acute toxicity of mepivacaine. J Pharm Pharmacol 40: 592-594

Bruguerolle B, Prat M (1988 b). Temporal variations of membrane permeability to local anaesthetic agents, bupivacaine and mepivacaine, documented by their erythrocytic passage in mice. Annu Rev Chronopharmacol 5: 227-230

Bruguerolle B. and Prat M. (1988 c). Chronotoxicity and chronokinetics of two local anaesthetic agents,bupivacaine and mepivacaine in mice. Annu Rev Chronopharmacol 5: 227-230

Bruguerolle B, Prat M (1990 a) Circadian phase dependent acute toxicity and pharmacokinetics of etidocaine in serum and brain of mice. J Pharm Pharmacol 42: 201-202

Bruguerolle B, Prat M (1990 b). Temporal variations in the erythrocyte permeability to bupivacaine, etidocaine and mepivacaine in mice. Life Sci 45: 2587-2590

Bruguerolle B, Prat M (1992). Circadian phase-dependent protein and tissular binding of three local anesthetics (bupivacaine, etidocaine and mepivacaine) in mice: a possible mechanism of their chronotoxicokinetics ? Chronobiol Int 9: 448- 452

Bruguerolle B, Valli M, Bouyard L, Jadot G, Bouyard P (1981). Circadian effect on carbamazepine kinetics in rat. Europ J Drug Metab Pharmacokinet 6: 189-193

Bruguerolle B, Valli M, Bouyard L, Jadot G, Bouyard P (1983). Effect of the hour of administration on the pharmacokinetics of lidocaine in the rat. Europ J Drug Metab Pharmacokinet 8: 233-238

Doran C, Catroux P, Bruguerolle B, Cambar J (1988) Circadian variations in mice erythrocytic membrane permeability to sodium and potassium. Annu Rev Chronopharmacol 5: 239-242

Focan C, Bruguerolle B, Arnaud C, Levi F, Mazy V, Focan-Henrard D, Bouvenot G (1988) Alteration of circadian time structure of plasma proteins inpatients with inflammation. Annu Rev Chronopharmacol 5: 21-24

Hecquet B, Meynadier J, Bonneterre J, Adenis L (1984) Circadian rhythm in cisplatin binding on plasma proteins Annu Rev Chronopharmacol 3: 115-118

Lemmer B, Bruguerolle B (1994) Chronopharmacokinetics. Are they clinically relevant Clin Pharmacokinet 26: 419-427

Lemmer B, Holle L (1991) Chronopharmacokinetics of imipramine and desipramine in rat forebrain and plasma after single and chronic treatment with imipramine. Chronobiol Int 8: 176-180

Lockard JS, Viswanathan CT, Levy RH (1985). Diurnal oscillations of CSF valproate in monkey. Life Sci 36: 1281-1285

Lutsch EF, Morris RW (1967) Circadian periodicity in susceptibility to lidocaine hydrochloride. Sciences 156: 100-102

Naranjo CA, Sellers EM, Giles HG, Abel JG (1980). Diurnal variations in plasma diazepam concentrations associated with reciprocal changes in free fraction. Brit J Clin Pharmacol 9: 265-272

Patel I, Venkataramanan R, Levy R, Viswanathan T, Ojemann LM (1982) Diurnal oscillations in plasma protein binding of valproic acid. Epilepsia 23: 282-290

Prat M, Bruguerolle B (1988) Temporal changes in the acute toxicity of mepivacainein mice. Med Sci Res 16: 715-716

Prat M, Bruguerolle B (1990 a) Circadian phase dependency of cardiac tissue levels of three amide type local anaesthetics. Annu Rev Chronopharmacol 7: 257-260

Prat M, Bruguerolle B (1990 b) Temporal variations of brain tissue levels of the three local anaesthetics in the mouse. Annu Rev Chronopharmacol 7: 261-264

Reinberg A, Reinberg MA (1977) Circadian changes of the duration of local anaesthetic agents. Naunyn-Schmiedeberg's Arch Pharmacol 297: 149-152

Reinberg A, Schuller E, Deslanerie N, Clench J, Helany M (1977) Rythmes circadiens et circannuels des leucocytes, proteines totales, IgA,G et M.Etude chez neuf sujets adultes jeunes et sains. Nouv Presse Med 6: 3819-3823

Reinberg A (1992) Concepts in chronopharmacology. Annu Rev Pharmacol Toxicol 32: 51-66

Riva R, Albani F, Ambrosetto G, Contin M, Cortelli P, Perucca È, Baruzzi A (1984) Diurnal fluctuations in free and total steady state plasma levels of carbamazepine and correlation with intermittent side effects. Epilepsia 25: 476-481

Valli M, Jadot G, Bruguerolle B, Bussiere H, Bouyard P (1979)Variations circadiennes des proteines plasmatiques chez le rat mâle adulte en synchronisation naturelle. J Physiol 75: 811-814

9 Photosynthesis: a Post-Transcriptionally Regulated Circadian Rhythm in Some Lower Eukaryotes.

T. Vanden Driessche
Département de Biologie moléculaire. Université Libre de Bruxelles. 46, av. des Ortolans. 1170 Bruxelles. Belgium.

Summary

Although not all Eukaryotes plants display a circadian rhythm in photosynthesis, most plants do. Much effort has been devoted to the understanding of the mechanism which regulates several higher plants photosynthesis CR· : it primarily results from a CR in nuclear coded mRNA level and synthesis rate of the light harvesting pigment-proteins complexes.

However, at least three unicellular eukaryotic organisms display a photosynthesis CR for which no transcription is required: Acetabularia, Euglena and Gonyaulax. Most evidently Acetabularia whose anucleate fragments maintain the CR up to 40 days. A brief summary of the evidence is given for all three species. In contrast,periodic translation on 80 S ribosomes appears to be needed for the CR expression - although the mechanism of control is unknown.

The experiments aimed at evidencing the processes which embody the photosynthetic circadian rhythm in the three investigated unicellular organisms are reviewed. The relevant reports pertain to both old and recent literature. The post-transcriptional photosynthesis CR control relies upon substructural organization changes.

In neither case could a CR in chlorophyll content or global a and b chlorophyll ratio be evidenced (or at least reliably related to the CR in photosynthesis). No

---

·Abbreviations: CR, circadian rhythm; CT, circadian time; DCMU, 3(3,4-dichlorophenyl) 1,1-dimethylurea; IAA, indolacetic acid; PPC, pigment-protein-complexe(s); PS I, photosystem I; PS II, photosystem II.

change in activity rate of the two individual photosystems could be detected in Acetabularia and Euglena.

In contrast, biophysical methods evidenced circadian changes. In Acetabularia, the Hill activity changes periodically in parallel with the O_2 evolution CR. Fluorescence emission spectra at 77° K revealed differences between the times of the day, likely due to changes in the molecular structure of the thylakoid membranes -interpretation strengthened by the spectrum produced after freeze-thaw treatments.

In Euglena, other methods were used. Emission spectra showed a consistent but low magnitude CR in energy transfer and large periodical oscillations in coordination between the two PS. The photosystems can be poised to the so-called "state I" (in which the PS I is excited) and, similarly, to the "state II" (in which the PS II is excited): analysis of low temperature emission spectra from chloroplasts in both states and of low temperature fluorescence and fluorescence ratio in the two states can only be be explained by changes in spill over between the PS.

More recent and sophisticated methods were used in analyzing the photosynthesis CR of Gonyaulax. It was found that the supramolecular organization of the pigment-protein-complexes changes periodically.

These organisms provide a demonstration of the power of membranes to generate circadian rhythmicity (possibly together with a cytoplasmic protein).

9.1 Introduction.

Photosynthesis circadian rhythmicity (CR) is more and more actively studied in higher plants.

From early reports it appeared that species, belonging to a variety of taxa, express circadian changes in O_2 evolution rate. The changes could be traced to CR in activity or content of enzymes or pigments or a combination of several such changes (reviewed by Vanden Driessche, 1984).

The important technical improvements that occured since that time allow to draw a clear distinction between transcriptionally regulated and post-transcriptionally regulated mechanisms of photosynthesis CR.

On the one hand, three unicellular organisms (i.e. those to which the question has been addressed) display a photosynthesis circadian rhythm (CR) for which transcription is dispensable. They are Acetabularia, Euglena and Gonyaulax, and their photosynthetic CR was studied from the mid-70ies to the 80ies. Only Gonyaulax photosynthesis CR has been re-examined in 1990 with new methodologies. How this transcription independent CR is physically embodied is the focus of the chapter. Its independence from transcription and probable dependence on translation are briefly summarized.

On the other hand, higher plants display a photosynthesis CR for which transcription is required. This type of CR regulation contrasts with the one cited above. Much recent research is devoted to the subject in a number of Mono- and

Dicotyledonous plants . Briefly, CR are controlled by every-day transcription primarily of the cab genes, a family of genes encoding the chlorophyll a/b binding proteins (see Meyer et al., 1989 and Kay and Millar, 1993 for reviews). In some species, but not in others, the level of nuclear mRNAs of several chloroplastic enzymes were found to oscillate (Kloppstech, 1985, Piechulla, 1988; see also the table of "clock genes" by Mc Clung, 1993). Kloppstech and collaborators (Adamska et al., 1991) showed the correlation between the oscillations of nuclear-coded mRNA level for three light-inducible chloroplast proteins (the light-harvesting chlorophyll a/b protein, the early light-inducible protein and Fe-S Rieske), their level in the integrated mRNA into polysomal complexes and in the protein level. Upstream cis-acting elements were identified that mediate circadian transcription of cab gene(s) in wheat and Arabidopsis (Nagy et al., 1988, Fejes et al., 1990, Millar and Kay, 1991 and the review concerning the molecular analysis of Arabidopsis CR by McClung, 1993). Some of the cab genes are also regulated by phytochrome. Further research will reveal exactly how the circadian regulation is operated at molecular level. In addition, new CR regulated chloroplastic enzymes will undoubtedly be described. The light-induced CR in $NADP^+$-glyceraldehyde-3-phosphate dehydrogenase for example was described in corn by Lillo in 1993.

It should be underlined that there is no reason to believe that a single CR controls the oscillation in photosynthesis (Adamska et al., 1991). In a C_3 Monocotyledonous plant, wheat (Triticum aestivum), experiments carried out by Sayeed and Mohanty (1988), described rhythmic oscillations in photosynthesis electron transport, in PS II and PS I activity, with a period of about 12-16 h in LL as well as in LD conditions. The peaks are expressed in L and D periods (or corresponding subjective periods). The PS II and PS I peaks alternate with each other. The ratio between the rates of the PS (PS I/ PS II) oscillates with a pattern phased rather like the PS II. The authors suggested that a temporal modulation of the photochemical activity (and the whole electron transport chain) is brought about by reversible conformation changes in the thylakoid membrane. These oscillations are semi-circadian and structural changes might occur with ultra- or circadian periodicity, simultaneously with a transcriptional CR.

9.2 Circadian Rhythmicity Regulated at Post-Transcriptional Level.

9.2.1 No Every-Day Transcription Is Required for the Photosynthesis CR Expression.

Acetabularia is a unicellular organism, and a giant cell of up to 5 cm long with characteristic morphology. It is remarkable in that it can be enucleated and survive for several weeks, carrying on its main physiological functions, including photosynthesis. This particular trait allowed the demonstration that, in Acetabularia, the CR in photosynthesis -the most easily and frequently monitored rhythm- keeps on unperturbed in anucleate algae: the actual presence of the nucleus is dispensable for the photosynthesis CR and, more generally for the CR of the alga. This has been explained by the fact that nuclear coded mRNAs, responsible for CR, are long lived and, additionally, stabilized in anucleate algae (Vanden Driessche, 1966 and 1967). The system has been extensively studied in order to try and understand the underlying mechanism of the photosynthesis CR in Acetabularia, using notably, an array of inhibitors (see Berger et al., 1987 and Vanden Driessche, 1994 for summaries).

Since both whole Acetabularia and anucleate fragments can be synchronized, it is of interest to note that the implicated light receptor is located in the plasma membrane, the chloroplast pigments exerting no effect (see Vanden Driessche, 1994 for a discussion).

In Gonyaulax, actinomycin D, an inhibitor of DNA transcription, abolishes the photosynthesis CR after two cycles at the high concentration of 2 $\mu g.ml^{-1}$. (ten times the "classical" concentration)(Karakashian and Hastings, 1962). Other inhibitors of the same process exert almost no effect (Karakashian and Hastings, 1963). Moreover, if the photosynthesis CR is comparable to the luminescence one and since no change in specific luciferin binding protein mRNA level could be detected in the 24 h cycle (Morse et al., 1990), one can speculate that transcription is not required either.

This point was not examined in Euglena.

9.2.2 A Role for Periodic Translation?

The periodic synthesis of at least one protein of unknown function is predicted on the grounds that inhibitors of protein synthesis carried out on 80 S ribosomes (1) lenghten the period of the rhythm and (2) induce phase-shifts when applied as pulses.

A number of experiments were carried out in Acetabularia with inhibitors acting on 80 S ribosomes: all affected the photosynthesis CR. In contrast, inhibitors acting on 70 S ribosomes did not. An extensive study of the effects of cycloheximide pulses resulted in a phase response curve with only delays when 8 h pulses were given from 23.9 to 7.7 h of the circadian cycle (Karakashian and Schweiger, 1976). The phase shift occurs in anucleate fragments as well, although the peak occurs 4 h earlier (Hartwig et al., 1986). Schweiger and his collaborators proposed that the control of the photosynthesis CR in Acetabularia could be regulated by a cytoplasmic protein P 230 (its apparent molecular weight is about 230 000). P 230 exhibits a pronounced CR in abundance as measured on scanning-densitometry profiles of ^{35}S-labelled protein fractions. The peak occurs at approximately CT 1. The phase is delayed by cycloheximide and the shift is the same as that produced on the photosynthesis CR (Hartwig et al., 1985).

The effects of 80 S ribosomes protein synthesis inhibitors was exhaustively studied by Hastings' group on the CR of luminescence of Gonyaulax. Rensing et al. (1980) found that the phase shift produced by cycloheximide on chloroplast ultrastructure corresponded to that predicted by the results obtained on luminescence. As a general rule, it can be assumed that the effects on the different CR are similar - even though the rhythms themselves may be differently phased-.

For Euglena, the typical effects of cycloheximide concentration were demonstrated on the phototaxism CR, producing a concentration-dependent period lenthening (Feldman, 1967).

In conclusion, the photosynthesis CR (as other CR) is affected by cytoplasmic ribosomal protein synthesis. In addition, in Acetabularia a putative protein for playing a role in the photosynthesis CR control has been isolated. Its function however remains speculative.

Besides change in protein synthesis, other means exist by which proteins can be modified around the clock. Phosphorylation is the one for which some experimental evidence has been obtained in Gonyaulax (Schröder-Lorenz and Rensing, 1987, Comolli et al., 1994).

9.3 The Support of the Circadian Rhythm in Photosynthesis in Three Unicellular Organisms.

9.3.1. The General Approach

In Acetabularia, Euglena and Gonyaulax, the molecules embodying the CR of photosynthesis have been investigated already twenty years ago. Although indirect, the early results yielded clear-cut results. More recently, the problem has been re-evaluated in Gonyaulax with modern techniques. The results stand in line with the former ones and ended up with the identification of the way in which molecules support the rhythm. In the three organisms, the possible correlation between ultrastructure and level of photosynthetic activity was first examined. Then, the strategy has been to reduce the photosynthetic system successively in order to determine which was the minimum one required for CR expression.

9.3.2 Dissecting the Photosynthesis CR

Ultrastructure (EM Studies). Acetabularia chloroplasts display remakable changes in ultrastructure according to time in the 24 h cycle. They involve the aspect and position of the thylakoids within the chloroplast, the number and volume of the polysaccharide granules and the appearence of the stroma, in LD as well as in LL (Vanden Driessche and Hars,1972a). The CR in ultrastructure, retained in anucleate algae, show some differences however when compared with that of whole cell chloroplasts (Vanden Driessche and Hars, 1972b).

An interesting point concerns a much more reduced anucleate system, that of cytoplasts. They are spontaneously released from algae surgically halved in a slightly hypotonic medium. They are droplets of cytoplasm ranging from 50 to 100 μm in diameter, thus variable in size but very fragile. Some of them are thin layered and contain a vacuole and surrounding cytoplasm with its chloroplasts. Other cytoplasts are devoid of vacuole and their density is much higher. They survive at best 11-12 days. In 3-days old cytoplasts, the CR of ultrastructure is still detectable, but no more in 11-days old ones (Vanden Driessche et al., 1973).

All CR in Acetabularia are synchronized by blue light and they keep on in constant darkness (DD) as well as in constant light (LL, in which case CR can be observed only on single cells because the oscillations become rapidly desynchronized, in even a small sample of algae (Karakashian and Schweiger, 1976).

In addition, three chloroplast populations can be distinguished in Acetabularia: the apical one consisting of young chloroplasts, with dense stroma and little polysaccharide material, mature one in the long stalk, with well developed thylakoids and larger polysaccharide granules and finally rhizoidal chloroplasts, filled up with reserve material. These developmental differences are superimposed to the CR. The photosynthesis CR differs in level of activity and parameters of the rhythm depending on the population age: the Hill reaction displays a higher base line of activity but a lower amplitude whereas the converse holds true for the amplitude (Vanden Driesche, 1974). Similar changes are observed for polysaccharide synthesis (Glory ,1979).

The question as to whether or not ultrastructure may reflect the photosynthetic activity has been examined in details in a variety of conditions. Except for the formation of pseudo-grana, always linked to darkness and reduced activity, there is no correlation between chloroplast ultrastructure and rate of physiological activity (Vanden Driessche and Hars, 1973).

In Euglena, some changes in chloroplast ultrastructure were also evidenced, mainly in the pyrenoids which were observed in the light period only. They appear to be transversed by thylakoids. No pyrenoid was present in the dark period (Könitz, 1965). Change in ultrastructure appears to be related with reserve material, a direct product of photosynthesis, rather than with photosynthesis itself.

In Gonyaulax, the first studies were essentially qualitative (Herman and Sweeney, 1975). Rensing and collaborators re-examined the problem and quantified the differences (Rensing et al., 1980). They confirmed the change in position of the chloroplasts within the cell, which leaves the central area devoid of chloroplasts at night or subjective night but penetrate the zone during the light or sujective light period (this part of the chloroplasts is qualified "expanded"). In this part -thus during the light period- there is generally a 2 thylakoids stacking whereas, in the outer, more peripheral region of the chloroplasts, thylakoid stacking displays a minimum number of 3-thylakoids during the first part of the light period and a maximum at night (or during the subjective night).The authors noted that no CR was observed in the course of the 24 h period of time in the interthylakoid distance in the distal part of the chloroplasts; however, the chloroplasts in their expanded region display larger distances between thylakoids than in the outer one.

Thus, for all three organisms, Acetabularia, Euglena and Gonyaulax, ultrastructure changes are observed at the level of the EM, but they do not support circadian changes in photosynthesis activity level except for one point: stacking correlates with low photosynthetic activity at least in Acetabularia and Gonyaulax (there is no information concerning Euglena).

CR in Chloroplast Fatty Acid Composition. Chloroplast fatty acid composition was examined in Acetabularia and a CR was evidenced (Jerebzoff and Vanden Driessche, 1983). Chloroplasts display a very large accumulation of palmitic acid at CT 9 h (3h after the photosynthesis CR peak) and lower peaks in oleic acid and C 18 polyunsaturated fatty acids at CT 6 in both LD and LL conditions. Whatever their location, within the thylakoid membrane, other chloroplast membrane or stroma, they might influence photosynthetic activity.

Only Acetabularia has been examined so far as concerns fatty acids CR

Photosynthesis as a Function of Irradiance at Two Different Times in the 24 h Cycle Another means of investigating the state of the thylakoids is to measure O_2 evolution rate as a function of light intensity (which gives the irradiance curve) at different times or conditions The differences between curves reflect differences in the thylakoid state. Duysens made use the method in the fifties and it was also found useful in estimating the photosynthetic apparatus performance in sun and shade leaves (Lichtenthaler, 1981).

In Acetabularia, the curves obtained in six experiments carried out in the middle of the light and dark periods display striking differences, the slope of the O2 evolution as a function of light intensity, being much steeper at the time of maximum photosynthesis than at minimum (fig. 9.1, unpublished representative experiment).

In Euglena, the experiments were also carried out at two different times in LL and transformed in linear plots. The plots show differences consistent with the highest values corresponding to the maximum photosynthesis rate (Lonergan and

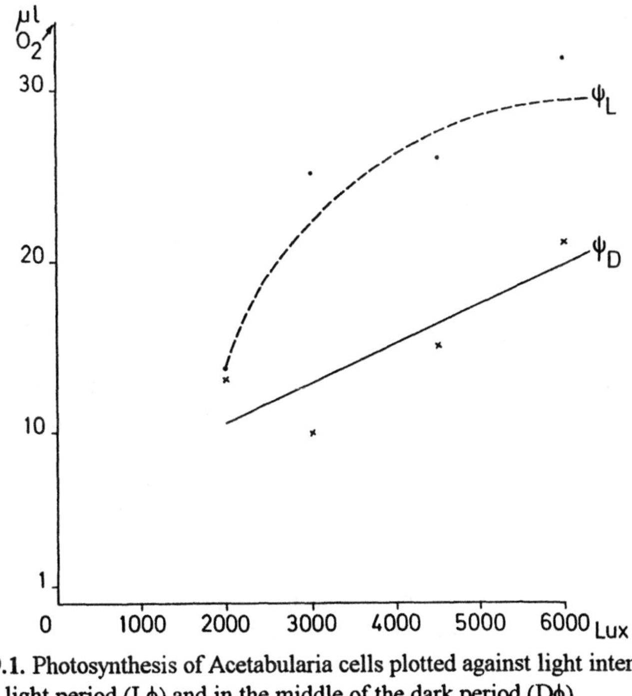

Fig. 9.1. Photosynthesis of Acetabularia cells plotted against light intensity in the middle of the light period (Lϕ) and in the middle of the dark period (Dϕ)

Sargent, 1979). The difference however decreased from the second day and the authors concluded that it indicates little change in the relative quantum yield. It is possible, however, that the CR was damping in their LL conditions.

Although the conclusions for both species do not appear sufficiently supported by statistics, the coherent results may be taken as an argument in favour of some kind of change in the thylakoid organization, resulting in changes in photosynthetic yield.

In Gonyaulax, photosynthesis-irradiance curves were also drawn (Prézelin and Sweeney, 1977). There is no statistical difference that could be related to the photosynthesis CR in this species, but the initial slope of the curve is circadian time-dependent and the changes parallel those in photosynthesis rate.

The Hill Reaction. The Hill reaction is a measure of the light-induced electron flow trough the whole electron chain: it is quantified by the rate of dichlorophenol indophenol (DCIP) reduction in the presence of an illuminated chloroplast suspension. The Hill reaction, with Acetabularia chloroplasts is rhythmic. It parallels the CR in O_2 evolution in the course of the day and is observed in young apical and in mature stalk chloroplasts (Vanden Driessche, 1974).

In Euglena, the highest value of the Hill reaction was also coincident with the the highest photosynthesis rate. The electron flow was measured by the rate of methyl viologen reduction by oxygen in the presence of the uncoupler, gramicidin D (Lonergan and Sargent, 1979, see also Lonergan, 1984).

Thus, in both species, the electron flow through the entire electron chain is clearly rhythmic.

Gonyaulax displays a CR in the Hill reaction although the electron flow through the PS I, including or not the plastoquinone pool, was constant with time (Samuelson et al., 1983).

The Chlorophyll Content and the a/b Chlorophyll Ratio. In Acetabularia, there is no change in the course of the day in chlorophyll concentration (Hellebust et al., 1967). This invariability was confirmed by Vanden Driessche et al. They also showed that no change in the ratio between chlorophyll a and b could be detected (Vanden Driessche et al., 1976).

In Euglena, similar results were obtained by Lonergan and Sargent (1978). Edmunds (1988) found some changes in chlorophyll content in LD but since they were either circadian or ultradian, whereas O_2 evolution remained circadian, no simple causality could be invoked between the two. In addition, the amplitude of the variations in chlorophyll content were too low to possibly account for the CR in photosynthesis.

Glenodium rather than Gonyaulax was studied for chlorophyll content. It is a marine dinoflagellate as is Gonyaulax. No CR could be detected in chlorophyll content (Prézelin and Matlick, 1980, cited by Lonergan, 1984).

The Photosynthetic Unit Size It is a measure of the number of chlorophyll molecules participating in the transfer of one electron from H_2O to $K_3Fe(CN)_6$. It is unlikely to be responsible for the photosynthesis CR: it was found to be unvariable in the course of the daily cycle (Lonergan and Sargent, 1979)

The Enzymes of the Calvin Cycle Since the availability of these enzymes may limit photosynthesis activity, it was important to examine whether or not they display circadian changes.

There is no change in activity rate of the 9 major enzymes of the Calvin cycle in Acetabularia (Hellebust et al. 1967).

In Euglena, ribulose bisphosphate carboxylase/oxygenase fluctuates in activity but neither sufficiently nor consistently to generate the CR (Edmunds, 1988). Two other enzymes were studied by Lonergan and Sargent(1978): glyceraldehyde-3-phosphate dehydrogenase, which remains constant throughout the 24 h cycle and carbonic anhydrase which is rhythmic only in LD conditions.

In Gonyaulax, contrary to earlier reports, no CR in activity was demonstrated for ribulose bisphosphate carboxylase-oxygenase (Sweeney, 1969, Bush and Sweeney, 1972).

The Activity of Invidual Photosystems. The next step aiming at determining the site of the circadian modulation responsible for the photosynthesis CR was to examine if the activity of the individual photosystems I or II varies with time in the 24 h cycle (Vanden Driessche and Lannoye, 1982).

As concerns the PS I, Acetabularia proved to be a difficult material since chloroplasts are very resistant to rupture, specially at certain periods of the year (presumably due to a seasonal rhythm) and yielding irregular results. Thus only experiments with intact chloroplasts could be considered. The PS I activity was very low and no circadian change could be detected in the course of the daily cycle. We reasoned that if such change occurs, it could be amplified by auxin (IAA) or its antagonist morphactin: both act on the photosynthesis CR (Vanden Driessche, 1979) and on the Hill reaction (Vanden Driessche, 1977). The results show a trend towards an IAA-induced enhancement of the activity of the PS I and a morphactin-induced inhibition but there is no time-dependent effect of either IAA or morphactin. As concerns the PS II, both substances dramatically affect its activity, IAA increasing and morphactin decreasing it. No differential effect with time in the 24 h cycle could be detected. Thus these results suggest that the photosynthesis CR of Acetabularia does not rely upon changes in the activity of individual PS.

Fluorescence Experiments The CR in Hill reaction and the absence of rhythmicity in individual PS activity suggest that circadian changes might occur in the coupling between PS. In order to test this hypothesis, a biophysical investigation on thylakoid structure was carried out: fluorescence emission spectra of algae frozen in liquid nitrogen were recorded at 77°K (Vanden Driessche et al. 1976).

Differences were observed depending on the time of the day in LD, DD and LL. Since there in no change in the amount of chlorophyll nor in the ratio between chlorophyll a and b, the differences could be accounted for in two different alternative ways: either a small change -even a very small one- in partitioning of chlorophylls could change both the photosynthesis activity and the fluorescence spectrum. Or it could be traced to the organization of the chlorophyll-protein complexes within the membrane in such a way as to modify the relative amount of energy transferred from the antenna (the energy collecting complexes) to the PS I or PS II centers. In order to discriminate between the two possibilities, the algae were submitted to successive, standardized "freeze-thaw" treatments. The differences between low temperature fluorescence emission spectra at different circadian times were increased. This reflects a change in the lability of the thylakoid organization. Consequently, the conclusion is that changes occur in the organization within the thylakoid membrane. These changes are likely to embody the CR in photosynthesis.

Euglena also displays no circadian change in the activity of individual PS. Lonergan and Sargent (1979) measured the whole cell fluorescence. It is attributable primarily to the chlorophyll a of the PS II. The most important results were the fluorescence kinetics obtained at 77° K after preillumination at 650 and 710 nm. Illumination of whole cells with 710 nm light excites primarily the PS I and results in chloroplasts poised in "state I". Similarly, illumination with 650 nm light excites primarily the PS II and results in chloroplasts poised in" state II". Both in states I and II, cells display a low amplitude CR in low temperature fluorescence (F). However, there is a high amplitude CR in the ratios F720/F650. These results suggest a CR in spill-over. The hypothesis is strengthened by the periodic effect of Mg^{2+} on chlorophyll fluorescence (Mg^{2+} induces a transfer of excitons).

Thus, both Acetabularia and Euglena display a CR in photosynthesis which is modulated at the level of the electron transport involving the two PS and their coupling. In other words, the CR in photosynthesis relies on the thylakoid membrane organization.

In Gonyaulax, not only was the intensity of chlorophyll a fluorescence higher during the light or subjective light part of the 24 h cycle than 12 h later and positively correlated with the rate of O_2 evolution (Sweeney et al., 1979), but the delayed light emission curves transients were also rhythmic -measured in a "long term", (3 to 4 min)- (Sweeney et al.,1982). The CR in fluorescence intensity of chlorophyll a persisted in the presence of DCMU, which blocks the electron flow between the two PS. This is consistent with the conclusion that the PS II (to which chlorophyll a is mainly associated) has a major role to play in the operation of the CR. The fact that the circadian changes in fluorescence intensity were abolished at 77° K (Govindjee et al., 1979) suggests that structural changes in the thylakoid membrane are critical in circadian rhythmicity.

In conclusion (1) there are some results in Gonyaulax which stand at variance with those obtained with Acetabularia and Euglena, the low temperature fluorescence for example and the circadian coupling of the two PS. (2) There is, however, a main common feature for all three organisms: the photosynthesis CR depends on thylakoid structural organization.

9.3.3 The Molecular Thylakoid Organization in Gonyaulax is Responsible for the CR in Photosynthesis: Gonyaulax.

In 1990, Knoetzel and Rensing (1990 a and b) took a molecular approach to the problem with Gonyaulax. They separated the five intrinsic pigment-protein-complexes (PPC) at 4 different times in the 24 h cycle and demonstrated that there is no change in pigment pattern. There is neither changes in free pigment amount after electrophoresis, in the middle of the light and dark periods.

In contrast, what does change with time is the pigment pattern of individual pigment-proteins, both in LD and in LL (fig.9.2). The main results of their analyses are summarized in Table 1. These changes in chlorophyll and carotenoid composition may be explained by changes in the organization of the pigments. A way of checking this hypothesis was to measure fluorescence induction kinetics in light favouring the peridin-containing antenna complexes and green light to excite pigments more closely associated with the PS II reaction center. The kinetics were biphasic, composed of a fast α-component and a slow β one. The authors interpreted the results by assuming two PS II forms (α and β). Their relative proportions changed with time in both LL and LD. In other words, there are changes in the supramolecular organization of some PPC according to time in the 24 h cycle. These changes could result from modifications in the organization of the PPC in the thylakoids.

The hypothesis was further supported by measurements of fluorescence induction of dark-adapted, DCMU-treated cells.

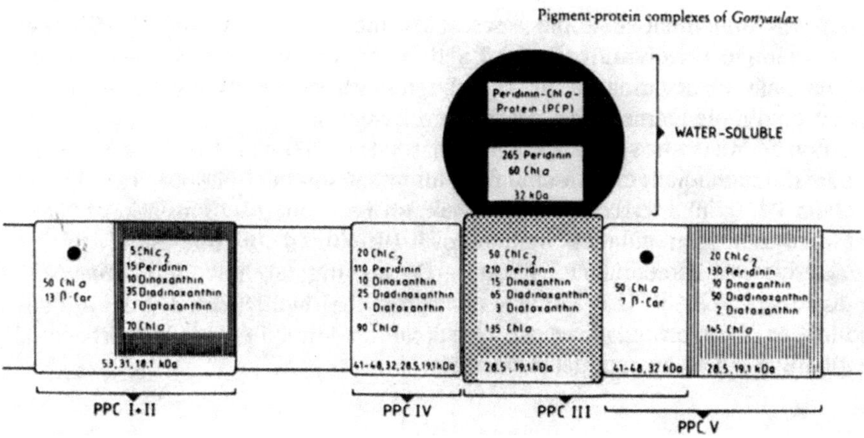

Fig. 9.2. Knoetzel and Rensing's model (1990a) for the organization of the pigment-protein complexes of Gonyaulax. (by permission)

Table 9.1. Analysis of the pigment-protein complexes of Gonyaulax (Knoetzel and Rensing, 1990b). The changes indicated refer to D or sujective D as compared to L or subjective L (D/L); ct, constant; xan = xanthophylls; chl a = chlorophyll a)

		LL conditions	LD
PS I	PPC I	ct L/D	
PS I	PPC I + II		xan/chla higher D/L
PS I	PPC II	ct L/D	idem
PS II	PPC III + IV	xan/chla higher in D/L	
PS II	PPC V	idem	ct L/D

9.4 Final Remarks and Conclusions

The three organisms, Acetabularia, Euglena and Gonyaulax display post-transcriptional regulation of the photosynthesis CR as single rhythm mechanism and they provide a remarkable case of circadian rhythmicity which does not depend on every day transcription. It could be that this type of photosynthesis circadian regulation is a characteristic feature of Protoctista-a hypothesis that we recently proposed (Vanden Driessche, 1994).It gives further weight to Sayeed and Mohanty proposal that at least part of the CR in wheat PS II activity is conformational.

The thylakoid membranes are operating the CR -this point is of major interest for our view that membranes may be crucial in the functioning of some CR: it is embodied by membranar conformation changes. There are differences however, between rhythmicity of some processes in the three organisms, the Hill reaction for example (involving both the PS I and II) or low temperature fluorescence. One possible explanation is that photosynthesis or its circadian regulation appeared more than once in the course of evolution.

Finally, there is strong evidence that a translational step takes place in the periodic (circadian) expression of photosynthesis in the three species. The protein could be P230 in Acetabularia. The link between this translational step and photosynthesis regulation is unknown. It would be interesting to further characterize Acetabularia P230 and investigate how it influences the photosynthesis CR. If P 230 actually affects the rhythm, could it exert a similar function in Gonyaulax whose thylakoid conformation would perhaps allow identifying a putative target molecule(s)?

Acknowledgments

We express our thanks to Prof. F. Homblé who critically read our manuscript and to Prof. E. Pays who gave us access to the facilities of the Laboratory.

References

Adamskà I,Scheel B,Kloppstech K(1991) Circadian oscillations of nuclear-encoded chloroplast proteins in pea (Pisum sativum . Plant Mol Biol 17:1055-1065

Berger S,de Groot EJ,Neuhaus G,Schweiger M(1987) Acetabularia: a giant single cell organism with valuable advantages for cell biology . Europ J Cell Biol 44:349-370

Bush KJ,Sweeney BM(1972) The activity of ribulose diphosphate carboxylase in extracts of Gonyaulax polyedra in the day and the night phases of the circadian rhythm of photosynthesis . Plant Physiol 50:446-451

Comolli J,Taylor W,Hastings JW(1994) An inhibitor of protein phosphorylation stops the circadian oscillator and blocks light-induced phase shifting in Gonyaulax polyedra . J Biol Rhythms 9:13-26

Edmunds LN Jr(1988) Cellular and molecular bases of Biological Clocks . Springer Verlag New York Berlin Heidelberg London Paris Tokyo . pp 497

Fejes E,Pay A,Kanevsky I,Szell M,Adam E,Kay S,Nagy F (1990) A 268 bp upstream sequence mediates the circadian clock regulated transcription of the wheat Cab-1 gene in transgenic plants . Plant molec Biol 15:921-932

Feldman JF(1967) Lenthening of the period of a biological clock in Euglena by cycloheximide, an inhibitor of protein synthesis . Proc Natl Acad Sci(USA)57:1454-1456

Glory M(1979) The circadian rhythm of polysaccharide synthesisin Acetabularia . Thesis, Université Libre de Bruxelles (Fac . Sciences)

Govindjee,Wong D,Prézelin BB,Sweeney BM(1979) Chlorophyll a fluorescence of Gonyaulax polyedra grown on a light-dark cycle and after transfer to constant light . Photochem Photobiol 30:405-411

Hartwig R,Schweiger M,Schweiger R,Schweiger HG(1985) Identification of a high molecular weight polypeptide that may be part of the circadian clockwork in Acetabularia . Proc Natl Acad Sci 82:6899-6902

Hartwig R,Schweiger R,Schweiger HG(1986) Circadian rhythm of the synthesis of a high molecular weight protein in anucleate cells of the green alga Acetabularia . Europ J Cell Biol 41:139-141

Hellebust JA,Terborgh J,McLeod GC(1967) The photosynthetic rhythm of Acetabularia crenulata II Measurements of photoassimilation of carbon dioxide and the activities of enzymes of the reductive pentose cycle . Biol Bull :133:670-678

Herman EM,Sweeney BM(1975) Circadian rhythm of chloroplast ultrastructure in Gonyaulax polyedra Concentric organization around a central cluster of ribosomes . J Ultrastruct Res 50:347-354

Jerebzoff S,Vanden Driessche T(1983) Rythme circadien de la teneur en acides gras dans la cellule entière et dans les chloroplastes d'Acetabularia mediterranea . C R Acad Sciences Paris 296:319-322

Kay SA, Millar AJ(1993) Circadian-regulated cab gene transcriptionin higher plants In: Young MW(ed)Molecular genetics of Biological Rhythms . Marcel Dekker,New York,Basel,Hong Kong, pp 73-89

Karakashian MW,Hastings JW(1962) The effects of inhibitors on macromolecular biosynthesis upon the persistent rhythm of bioluminescence in Gonyaulax . J Gen Physiol:1-12

Karakashian MW,Hastings JW(1963) The inhibition of a biological clock by actinomycin D .Proc Natl Acad Sci USA:2130-2137

Karakashian MW,Schweiger HG(1976) Circadian properties of the rhythmic system in individual nucleated and enucleated cells of Acetabularia mediterranea . Exptl Cell Res 97:366-377

Kloppstech K(1985) Diurnal and circadian rhythmicity in the expression of light-induced plant nuclear messenger RNAs . TIBS 16:502-506
Knoetzel J,Rensing L(1990a) Characterization of the photosynthetic apparatus from the marine Dinoflagellate Gonyaulax polyedra I Pigment and polypeptide composition of the pigment-protein complexes . J Plant Physiol 136:271-279
Knoetzel J,Rensing L(1990b) Characterization of the photosyntheticapparatus from the marine Dinoflagettate Gonyaulax polyedraII Circadian rhythmicity of photosynthesis and the supromolecular organization of pigment-protein complexes . J Plant Physiol 136:280-288
Könitz W(1965) Elektronmikroskopishe Untersuchungen an Euglena gracilis im Tagesperiodischen Licht-Dunkel Wechsel . Planta 66:345-373
Lichtenthaler HK(1981) Adaptation of leaves and chloroplasts to high quanta fluence rates In:Photosynthesis VI Photosynthesis and Productivity, Photosynthesis and Environment . Akoyunoglou G 273-287
Lillo C(1993) Light-induced circadian rhythms in NADP+- glyceraldehyde-3-phosphatedehydrogenase mRNA in corn seedlings . J interdiscipl Cycle Research 24:65-71
Lonergan TA(1984) Regulation of the photosynthetic light reactions by the biological clock . Photochem Photobiol 44:789-793
Lonergan TA,Sargent ML(1978) Regulation of the photosynthesis rhythm in Euglena gracilis I Carbonic anhydrase and glyceraldehyde-3-phosphate dehydrogenase do not regulate the photosynthesis rhythm . Plant Physiol 61:150-153
Lonergan TA,Sargent ML(1979) Regulation of the photosynthesis rhythm in Euglena gracilis II Involvement of electron flow through both photosystems . Plant Physiol 64:99-103
McClung CR(1993) The higher plant Arabidopsis thaliana as a model system for the molecular analysis of circadian rhythms . In: Young MW(ed) Molecular genetics of Biological rhythms . Marcel Dekker,New York,Basel,Hong Kong, pp 1-35
Meyer H,Thienel U,Piechulla B(1989) Molecular characterizationof the diurnal/circadian expression of the chlorophyll a/b binding proteins in leaves of tomato and other dicotyledonous and monocotyledonous plant species . Planta 180:5-15
Millar AJ,Kay SA(1991) Circadian control of cab gene transcription and mRNA accumulation in Arabidopsis . The Plant cell 3:541-550
Morse DS,Fritz L,Hastings JW(1990) What is the Clock? Trends in Biological Science 15:262-265
Nagy F,Kay SA,Chua N-H(1988) A circadian clock regulates transcription of the cab-1 gene . Genes and Devel 2:376-382
Piechulla B(1988) Plastid and nuclear mRNA fluctuations in Tomato leaves - diurnal and circadian rhythms during extendeddark and light periods . Plant Molec Biol 11:345-353
Prézelin BB,Sweeney BM(1977) Characterization of photosynthetic rhythms in marine Dinoflagellates II Photosynthesis-irradiance curves and in vivo chlorophyll a fluorescence . Plant Physiol 60:388-392
Rensing L,Taylor W,Dunlap J,Hastings JW(1980) The effects of protein synthesis inhibitors on the Gonyaulax clock II the effect of cycloheximide on ultrastructural parameters . J comp Physiol 138:9-18
Samuelson G,Sweeney BM,Matlick HA,Prézelin BB(1983) Changes in Photosystem II account for the circadian rhythm in photosynthesis in Gonyaulax polyedra . Plant Physiol 73:329-331
Sayeed SA,Mohanty P(1988) Rhythmic oscillations in wheat chloroplast photochemical activity I Oscillations in whole chain photosystem II and photosystem I electron transport activities . Proc Indian Acad Sci 98:157-174

Schröder-Lorenz A,Rensing L(1987) Circadian changes in protein -synthesis rate and protein phosphorylation in cell-free extracts of Gonyaulax polyedra . Planta 170:7-13
Sweeney BM(1969) Transducing mechanisms between the circadianclock and overt rhythms in Gonyaulax . Can J Bot 47:299-308
Sweeney BM,Kazuhiko S,Fork DC(1982) A comparison of the fluorescence and delayed light emission in Gonyaulax polyedra in day and night phases of the circadian rhythm in constant light . Carnegie Year Book 81,68-70
Sweeney BM,Prézelin BB,Wong D and Govindjee(1979) In vivo chlorophyll a fluorescence transients and the circadian rhythm of photosynthesis in Gonyaulax polyedra . Photochem Photobiol 30:309-311
Vanden Driessche T(1966) The role of the nucleus in the circadian rhythms of Acetabularia mediterranea . Biochim Biophys Acta 126:456-470
Vanden Driessche T(1967) The nuclear control of the chloroplasts circadian rhythm Science Progress 55:293-303Vanden Driessche T(1974) Circadian rhythm in the Hill reaction of Acetabularia . In: Avron M(ed)Proc 3d Internat Congress in Photosynthesis Elsevier Scientific Publ Co, pp 745-751
Vanden Driessche T(1977) Regulation of circadian rhythmicity in Acetabularia: morphactins as a tool for determining the role of membranes . In:Scharf J-H and von Mayersbach(eds) . Nova Acta Leopoldina 225(44)pp293-299
Vanden Driessche T(1979) Phase-shifting effect of IAA on the photosynthesis circadian rhythm of Acetabularia . In:Bonotto S, Kefeli V,Puiseux-Dao S(eds) . Developmental Biology of Acetabularia . Elsevier/North Holland Biomed Press Amsterdam pp 195-204
Vanden Driessche T(1984) Photosynthesis, Circadian rhythms and Herbicides . In: Sybesma C (ed) Advances in Photosynthesis Research . Martinus Nijhoff/Dr W Junk publ IV 3 pp 301-304
Vanden Driessche T(1994) Circadian rhythms in three unicellular organisms . The pecularities of the organisms, the evidence they brought on rhythms and their specific practical problems . Outline of recent hypotheses . Biol Rhythm Research 25:345-386
Vanden Driessche T,Hars R(1972a) Variations circadiennes de l'ultrastructure des chloroplastes d'Acetabularia . I Algues entières . J Microscopie 15:85-90
Vanden Driessche T ,Hars R(1972b) Variations circadiennes del'ultrastructure des cloroplastes d'Acetabularia . II Algues anucléées . J Microscopie 15:91-98
Vanden Driessche T,Hars R(1973) Ultrastructure of the chloroplasts of Acetabularia mediterranea and rate of physiological activities . Arch Biol 84:539-551
Vanden Driessche T,Lannoye R(1982) Time-dependent effects of auxin and anti-auxin on photosynthesis in Acetabularia . Internat J Chronobiol 8:97-104
Vanden Driessche T, Dujardin E,Magnusson A,Sironval C(1976) Acetabularia mediterranea: circadian rhythm of photosynthesis and associated changes in molecular structure of the thylakoids . Internat J Chronobiol 4:111-124
Vanden Driessche T,Hars R,Hellin J,Boloukhère M(1973) The substructure of cytoplasts obtained from Acetabularia mediterranea . J Ultrastuct Research 42:479-490
Vanden Driessche T,Lateur L,Rzmieniewski,Guisset J-L(1988) Natural sea-water may be unreliable for the culture of Acetabularia . Chronobiol Internat 5:1-3
Vanden Driessche T,Guisset J-L,Gaspar T,Kevers C,Koukkari WL (1989) Masking in plants . Chronobiol Internat :13-19
9txx6160

10 Circadian Rhythms, Membranes and Susceptibility to Environmental Factors

E. Wagner[1], S. Kiefer[2], C. Penel[2], J. Normann[1], S. Ruiz-Fernandez[1], M. Bonzon[2] and H. Greppin[2]

[1] Biologisches Institut II, Universität Freiburg, Schänzlestr. 1, D-79104 Freiburg;

[2] Laboratoire de Physiologie végétable, Université de Genève, 3, Place de l'Université, CH-1211 Genève 4

Summary

Circadian rhythm in energy metabolism and circadian rhythmicity of plasmamembrane-linked functions are possibly closely coupled. Rhythmic stimulation of plasmamembrane-bound receptors could feed-back on cellular energy metabolism (phase shifting) which in turn could rephase sensitivity changes of plasmamembrane-bound receptors and metabolic function (e.g. transporters and channels).

Indications for membrane-associated rhythmic functions are deduced from rhythmic phytochrome pelletability in *Cucurbita,* the rhythmic binding of alkaline peroxidase to plasmamembrane microsomes from *Pharbitis* and the amplitude modulation of this rhythm by phytochrome photoreversibility, only at a particular phase of the rhythm.

These results clearly indicate an endogenous rhythmic control of the phase for potential amplitude modulation by the photoreceptor phytochrome in *Pharbitis,* a phenomenon earlier observed in *Chenopodium* for enzymes involved in energy metabolism.

The short-day plant *Chenopodium rubrum* is characterized by rhythmic functions in the different organs (leaves, shoot, root, apical meristem). Organization of activity in the whole plant seems to depend on the communication between the different organs and tissues via frequency-coded (electrical) signals.

A 20 h rhythm in cell division at the apex might impose rhythmic limits on the effectiveness of a floral stimulus originating in a 30 h rhythmic pattern from the leaves. A two frequency oscillation in the pattern of a circadian, rhythm in stem-extension-rate (SER) is considered to result from the interaction between rhythmic functions which originate in the leaf and in the stem or apex.

Changing the phase or frequency of the leaf-based oscillation by the critical photoperiod via phytochrome could result in a modified interaction between leaf- and shoot -based rhythmic (electrical) activity and might thus lead to a change in the rates and planes of cell division at the apex to allow flower initiation.

10.1 Introduction

Growth and differentiation in most living systems are tightly coupled to seasonal changes in thermo- and photoperiod. Photo- and thermoperiodic phenomena show a rhythmic change in sensitivity which is endogenous in character and reflects an endogenous circadian rhythm of metabolic activity. The network of cellular energy metabolism can thus be viewed as an evolutionary adaptation to a cyclic energy supply from the environment. It is assumed that in the course of evolution circadian metabolic activity became the innate timer in eukaryotes and some prokaryotes, controlling growth, differentiation and behaviour [Wagner 1976a,b,c, 1977, 1983; Wagner et al.1985, Wagner et al.1983].

By programming behaviour in relation to the seasonal change in the daily photoperiod, plants and animals adapt to predictable variations in the environment. By timing dormancy and initiation of flowers at a particular season, the continuity of species is assured. In such photoperiodically controlled responses, a time measuring system, an endogenous circadian rhythm, is implied. In plants, photoperiodic behaviour has been used very successfully to define operationally a photoreceptor system - phytochrome - which distinguishes between light and dark through a photoreversible process and the action of the active form of the pigment on metabolism. Metabolic activities of plants change drastically at dusk from dominant photosynthesis to utilization of reserves, and the reverse occurs at dawn. The change in the pigment and the timing in the availability of "substrates" on which it acts are supposedly the essentials of photoperiodic control. Circadian rhythmicity in energy metabolism thus seems to be a consequence of and necessity for activity regulation between subsystems of cells [Srere and Mosbach,1974]. In an attempt to elucidate the mechanism of endogenous rhythmicity and phytochrome action we have emphasized the importance of cellular energy metabolism [Wagner 1976a,b,c, 1977, Wagner et al. 1985a; Wagner and Cumming,1970; Wagner anf Fukshandy,1985; Wagner et al. 1983,1993].

Taking into account the symplastic organisation of higher plants [Lucas et al. 1993], the concept of compartmental feedback in energy metabolism on the cellular level becomes even more attractive in view of known bioelectric phenomena [Tanada,1983]. The symplast might be the basis for transduction of electric, photoperiodic and photomorphogenetic stimuli. The integration of activity of an entire plant could be based on the same symplastic organisation of the organism to give rise to a circadian rhythm in electric surface potentials [Aimi and Shibasaki, 1975; Davies, 1987a,b; Davies et al.,1991; Yamaguti, 1932]. The circadian rhythm in transcellular current of a single cell, as observed in *Acetabularia* [Novak and Sironval,1976] probably arises from a circadian rhythm in energy metabolism and rhythmicity of transport processes at the plasma membrane [Mayer and Fisher,1994; Mills et al.,1994; Pickard,1994.].

A circadian rhythm in energy metabolism might thus be reflected in a corresponding rhythm in the state of energy-tranducing biomembranes which could integrate the binding sites for the photoreceptors (phytochrome and/or flavoproteins). The interaction of photoreceptor and binding site could therefore be controlled by energy metabolism. Photostimulation of the receptor could lead to transitory modulation (phase setting) of oscillatory energy metabolism and switching of metabolic controls. The energy dependent state of membranes is in turn considered to determine the sensitivity of membrane-bound receptors and enzyme systems.

The experimental results support the concept that the different subsystems of energy metabolism are coordinated in space and time to avoid wasteful substrate cycles, i.e. futile cycles and/or to avoid (photo-) oxidative stress [Foyer et al.1994; Pahl and Baeuerle,1994; Paietta,1982]. The daily and seasonal adaptation of plants is correlated with characteristic changes in energy metabolism. The passage from germination to vegetative growth and reproduction is reflected in specific patterns of energy metabolism. This is demonstrated by systemic changes in redox and phosphorylation state and modulation of the enzymatic capacity of the cell subsystems. From the behaviour of the systems under constant conditions or under highly artificial environmental cycles it is concluded that diurnal (24 h) synchronization of energy metabolism is essential for survival. Deterioration of the specific phase-relationship between the cellular energy transduction subsystems will result in pathological development and cell death.

10.2 Experimental Evidence

10.2.1 Modulation of Compartmentalized Enzymes by Circadian Rhythm and Photoreceptor Interaction

In developing seedlings of *Chenopodium rubrum*, both the adenylate kinase (AK) and glyceraldehyde-3-phosphate dehydrogenase (GPD) capacities increase under phytochrome control [Frosch and Wagner,1973b]. With established green seedlings reciprocal semicircadian oscillations in glycolytic and photosynthetic GPD can be demonstrated [Frosch and Wagner,1973a]. With AK semicircadian and circadian oscillations in enzyme capacity have been observed [Frosch and Wagner,1973a]. In AK, as well as in photosynthetic GPD, the circadian rhythm predetermines the phase at which phytochrome can modulate the amplitude of the enzyme capacity oscillation. The involvement of compartmentalized isozymes in the oscillations of GPD has been shown, with evidence of pyridine nucleotide mediation of circadian and phytochrome control [De Looze and Wagner,1983a,b,c]. Changes in the metabolic and developmental state of seedlings are reflected in their AK pattern, and these correlate specifically with the transition from vegetative to reproductive development [Wagner et al.,1985; Wagner et al.1983].

In the case of AK multiple forms of the enzyme were found and could be respectively localized in mitochondria (AK I), cytoplasm (AK II), chloroplasts (AK III and V) and possibly the nucleus (AK IV) [Wagner et al.1983]. Like glycolytic GPD, most certainly all AK forms are synthesized on cytoplasmic ribosomes and hence, to reach other compartments have to be translocated there specifically

10.2.2 Considerations of Protein Translocation and Phytochrome Action

Phytochrome modulated membrane biogenesis in plants as deduced from photoperiodic control of stem extension rate and rhythmic growth movements of leaves (cf. 10.2.5) may involve so-called small GTP-binding proteins [Verma et al.1994]. It is proposed that a phytochrome-ribonucleoprotein interaction, as evidenced by changes in pelletability (cf. 10.2.3), might be significant for phytochrome action in vivo on membrane-organized translocator systems [Marmé,1977] e.g. for co-translational and/or post-translational translocation of proteins across membranes [Walter and Blobel,1981].

10.2.3 Circadian Rhythms and Phytochrome Binding to Ribonucleoprotein

In view of the possibility that membranes might be the sites for primary action of phytochrome [Marmé,1977] it is of basic importance to study binding phenomena of the pigment *in vivo*. In studying the association of phytochrome with pelletable structures [ribonucleoprotein, Quail,1975a,b] it was shown in *Cucurbita pepo* L. [Jabben and Schäfer,1976] that after a 12 h white light induction the total amount of phytochrome and its pelletability oscillate in the dark with frequency periods of 24, 12 and 6 h. The pelletable fraction oscillated between 20 and 60 %. The rhythmic oscillations were attributed to the white light induction period. Without this, the pelletable fraction remained constant at about 56 %. These experiments demonstrate a rhythmic change in the ability of pelletable structures to bind phytochrome to 'receptor sites'.

In another set of experiments [Horwitz and Epel,1978] with etiolated seedlings of *Cucurbita pepo* L. it was found that the spectral properties of active phytochrome (P_{fr}), as measured *in vivo*, exhibited a circadian rhythm with respect to peak position between 728 and 737 nm with a period of 22 - 24 h. In correlation with these changes it was found that the responsiveness of hook opening to a single red light irradiation also exhibited a circadian period. It was suggested that the long wavelength form of P_{fr} may correspond to a complex of P_{fr} with its native receptor and that it might be the change in the level of this complex which governs circadian responsiveness. In the case of phytochrome pelletability the white light period might have only amplified the response [c.f. Jabben and Schäfer,1976].

10.2.4 Photoperiodic and Phytochrome Control of Peroxidase Binding to Microsomes from *Pharbitis nil* Cotyledons

Pharbitis nil is a short-day plant which has been widely used in studies on photoperiodic control of flowering [Vince-Prue and Gressel,1985]. Peroxidases exist as numerous isozymes and these have been used as markers of specific gene function in metabolic regulation during adaptation to environmental changes and in response to stress conditions. Alkaline test-peroxidase was extracted from 8 d old cotyledons of seedlings grown in continuous light. The acidic isoperoxidases were removed by chromatography on DEAE-Sephacel, and the alkaline peroxidases used for the binding test.

Manganese-mediated binding of alkaline isoperoxidase from *Pharbitis nil* to microsomal membranes required the presence of the sugar moiety on the glycoprotein [Kiefer et al.1985]. Microsomal membranes from a crude preparation are characterized by marker enzymes [Sandelius et al.,1986]. It was anticipated that an endogenous rhythm in metabolic activity should result in rhythmic changes in the "make up" of membranes and that phytochrome should act via modulation of membrane properties. [Wagner et al.,1985; Wagner and Cumming,1970; Wagner et al.,1983].

Seedlings of *Pharbitis nil* are induced to flower by one short-day five days after sowing at their cotyledonary stage. The effect of the inductive night is reversibly inhibited at the seventh hour of darkness by red/far red light treatments acting via the phytochrome system.

Seeds of *Pharbitis nil* (cv. Violet) were germinated in moist vermiculite for 4 d at 20 °C in continuous white light. Thereafter they received a daily temperature programme of 8 h at 26 °C plus 16 h at 22 °C. From the second day of the temperature cycle they were kept in the dark during the 16 h at 22 °C, i.e. from 17.00 h to 9.00 h, with or without a 3 min red light pulse at the 7th h of darkness. (One group of plants received red/far-red light treatments to demonstrate phytochrome photoreversibility).

During the second day of the same photo-thermoperiodic conditions the following procedures were carried out:

a) Peroxidase-free microsomes [Kiefer et al.,1987] were prepared at three hour intervals from cotyledons of plants which had either received a red light pulse or had been kept in darkness and tested for manganese-dependent peroxidase binding. (Fig. 10.1).

b) Peroxidase-free microsomes were prepared at 18.00 h (second day) from cotyledons which had received the following treatments at the 7th h of darkness: none; 3 min red; 3 min far-red; 3 min red plus 3 min far-red. The manganese-dependent binding of alkaline test-peroxidase was examined (Fig. 10.2).

The results of Fig. 10.1 demonstrate a diurnal rhythm in the level of manganese-dependent peroxidase-binding to microsomes isolated at different times throughout a 8 h light : 16 h dark photoperiodic cycle. The level of binding

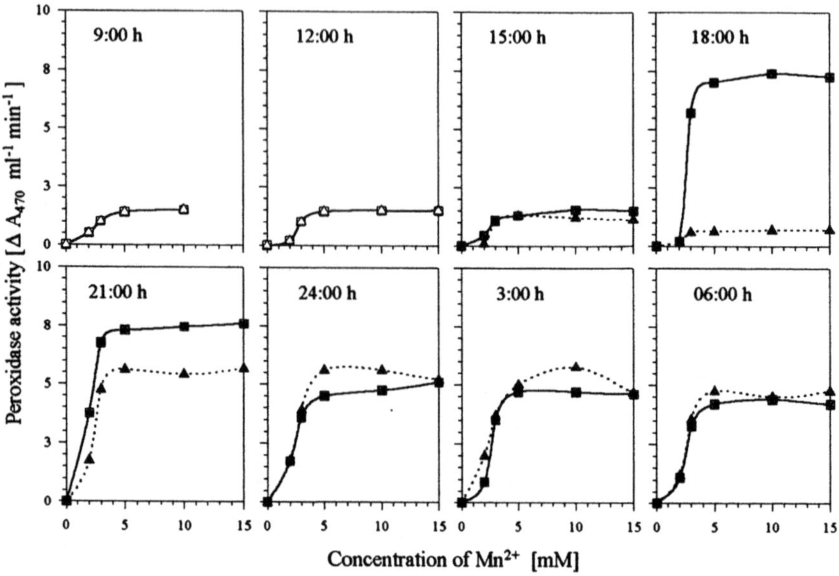

Fig. 10.1. Manganese-dependent binding of alkaline test-peroxidase to microsomes isolated from cotyledons at the hours given in the figure. (□), night interruption by red light pulse at the 7th h of darkness; (Δ), dark control without red light pulse

can be modulated by interrupting the night with a red light pulse. The red light effect is mediated by phytochrome as shown in Fig. 10.2. A phytochrome-modulated diurnal rhythm in the "make up" of microsomal membranes derived from the plasmalemma and the tonoplast is inferred.

10.2.5 Circadian Rhythms in Stem Extension Rate (SER) and Leaf Movements as Related to Photoperiodic Control of Flower Initiation

Photoperiodic control of flowering involves the following steps: the perception of the critical photoperiod by the leaves, the production and the transduction of a flowering signal from the leaves via the petioles and the stem axis to the apex to change the pattern of rates and planes of cell division at the apex in the process of floral evocation [Wagner and Fukshansky,1985].

An induced leaf sustains signal production, as can be demonstrated by grafting an induced leaf to a vegetative plant, with subsequent induction of flowering of the recipient plant soon after a living connection has developed. A successful graft union is correlated with a change in electrical resistance between stock and scion to the level of the intact system, indicating that symplastic connection has been completed [Yang et al.,1992].

It has been demonstrated that several rhythms in plants of *Chenopodium rubrum* L. influence floral induction [King,1975]. The photoperceptive leaves of *Chenopodium* produce a leaf-induced rhythmic process with a period length of

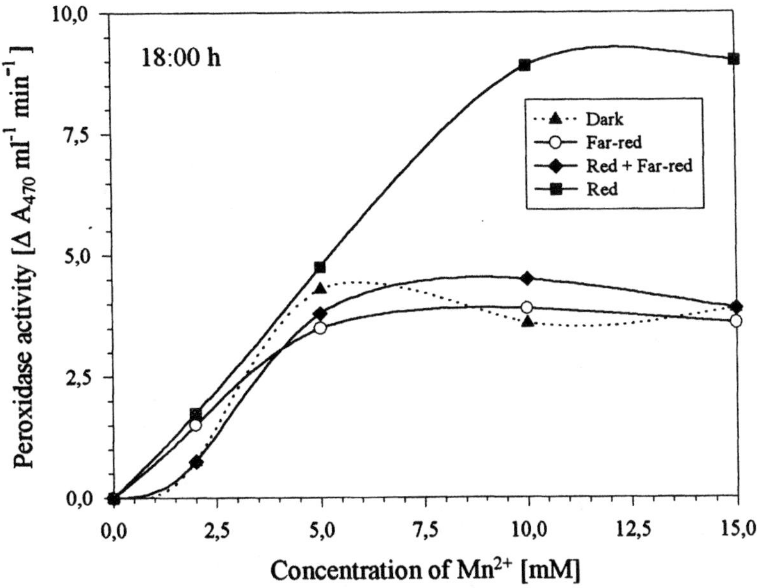

Fig. 10.2. Manganese-dependent binding of alkaline test-peroxidase to microsomes isolated from cotyledons which had received the light treatment indicated in the figure

30h. There is also a rhythmic 30-h periodicity in the flowering response to duration of darkness. In addition there is a rhythm with a period length of about 24 h which determines the response of the seedlings exposed to a fixed 12-h dark period at different times during the day. The 24-h and 30-h oscillations seem to be of distinct origin. A 20-h periodicity in the rate of cell divisions at the shoot apex might impose rhythmic limits on the effectiveness of the floral signal originating from the leaves and triggering floral evocation at the apex. A periodicity in sensitivity of floral development at the shoot apex is also inferred from the effectiveness of 6-h applications of solutions of glucose, ethanol, or gibberellic acid, which possibly affect the progression of the cells through the cell cycle [King,1975].

A detailed kinetic analysis of growth rate of *Chenopodium* stems has been used to study the timing of photoperiodic flower initiation and related rhythmic inter-organ communications [Ruiz-Fernandez and Wagner,1989,1994].

The stem extension rate (SER) of three ecotypes of the short-day plant *Chenopodium rubrum*, monitored with a fully automatic auxanometric system, displays all the features of circadian rhythms and responds immediately to photoperiodic treatments. In continuous light stem elongation is characterized by a circadian oscillation in growth (see Fig. 10.4). The three ecotypes exhibit distinct period lengths which correlate with their critical photoperiods and their geographical origin. Under daily light-dark cycles the stems grow with a diurnal rhythm, showing oscillations in the distribution of growth between light and dark spans (Fig. 10.3). From the SER kinetics the ratio of the integral growth during

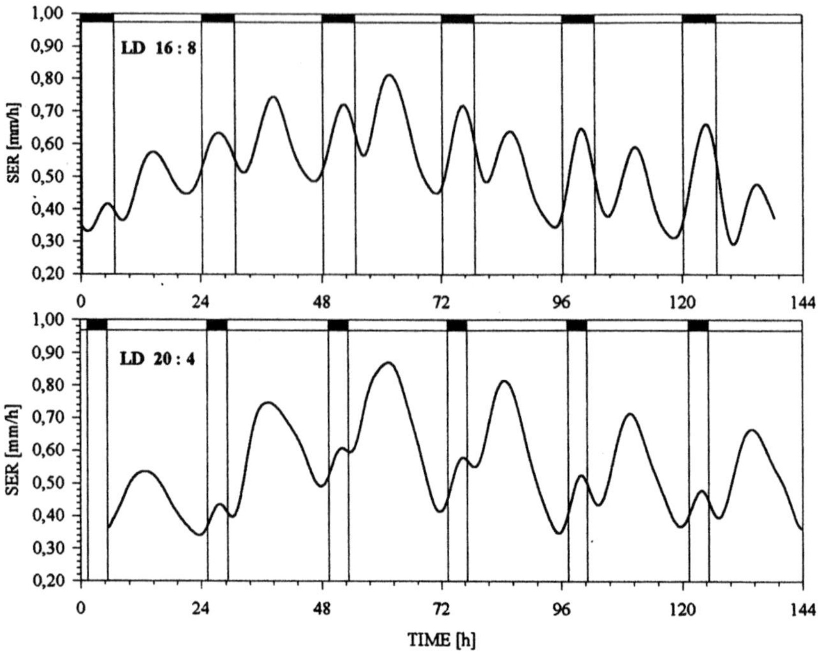

Fig. 10.3. Stem extension rate (SER) of *Chenopodium rubrum* under a flower-inducing light-dark cycle of 16 h light and 8 h darkness (upper graph) and a vegetative light-dark cycle of 20 h light and 4 h darkness (lower graph). Black bars at the top of each graph indicate the dark periods

the dark span over the integral growth during the light span (DLGR) was calculated. Under all photoperiods tested so far, floral induction required a DLGR-threshold value of 0.6 in all cases. Plants showing a DLGR-value below 0.6 always remain vegetative. This held true for all ecotypes, such that a DLGR-threshold value of 0.6 coincided with the critical photoperiod of each ecotype.

Under continuous blue light a two-frequency oscillation in the pattern of the circadian rhythm in SER is detectable. This observation can be explained by the existence of two endogenous oscillators. The higher frequency oscillation under these conditions corresponds to the oscillation frequency of defoliated plants. Thus this oscillation is supposed to be located in the stem or apex.

On the other hand, intact flowering plants, in contrast to vegetative plants, show a significantly shorter period length in SER. This phenomenon cannot be observed with flowering defoliated plants. Since it is well established that the flower inducing "principle" is produced exclusively in the leaves, we concluded that the lower frequency oscillation might originate in the leaves. Thus the resulting SER-pattern could be considered as an interaction between two oscillations, one in the leaf and the other in the stem or apex. Changing the frequency of the leaf-based oscillation by an appropriate environmental signal [Bünning,1977] would then result in a modified interaction between the oscillations (leaf and shoot) and/or phase relationship between the rhythm in SER

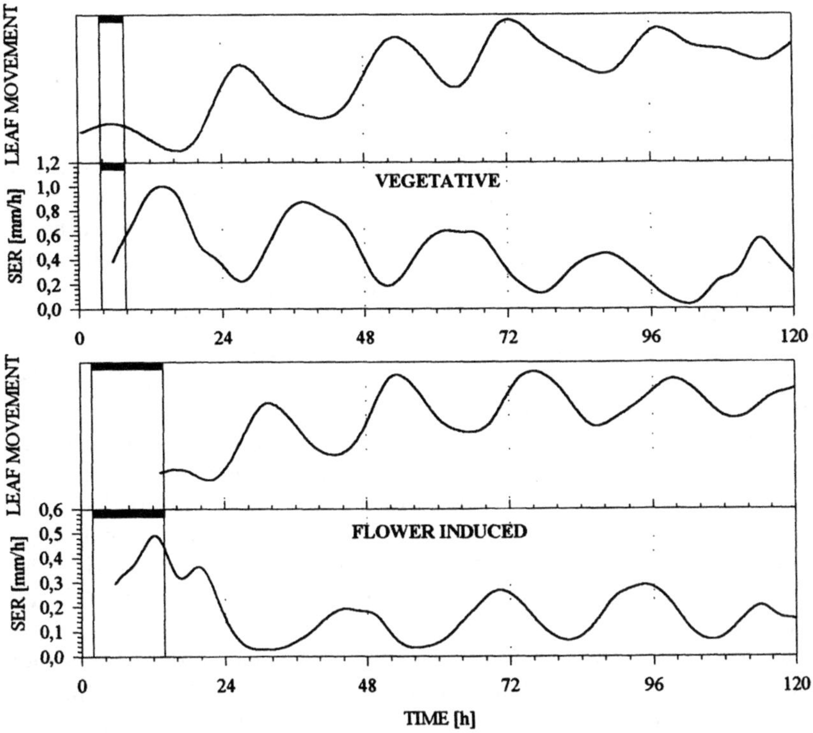

Fig. 10.4. Stem extension rate (SER) and leaf movement of *Chenopodium rubrum* in constant light after a vegetative light-dark cycle of 20 h light and 4 h darkness (upper graph) and a flower-inducing light-dark cycle of 12 h light and 12 h darkness (lower graph). Black bars at the top of each graph indicate the dark periods.

and leaf movement (Fig.10.4) and possibly could provide the information needed to induce flower formation in the apex [Fukshansky,1981].

Induction of flowering and defoliation significantly shortens the period length of the circadian rhythm in SER. From these results we conclude that the period length of the circadian rhythm in SER is determined by an interaction between leaves and stem and that this interaction is changed upon perception of the critical photoperiod by the leaves. The flower-inducing state of a leaf seems to be very stable, as can be deduced from grafting experiments in different plants as mentioned above. The grafting experiments were taken to support the hypothesis on the involvement of a flowering hormone. Rhythmic interactions between plant tissues and organs might be essential for control of growth and differentiation via environmental signals [Bertram and Karlsen,1994; Millet et al.,1988] based on the circadian rhythmic behaviour of single cells [Mayer and Fischer,1994] and might be coordinated by the symplastic organization of the entire plant [Lucas et al.,1993].

10.3 Conclusions

In view of demonstrated rhythmic and photoperiod mediated phenomena in surface potentials [Aimi and Shibasaki,1975; Davies,1987; Davies et al.,1991; Yamaguti,1932], we suggested [Wagner, et al.,1985] that the flowering stimulus might be an electrical or electro-chemical signal. A similar concept was advanced in relation to systemic effects from wound responses [Davies,1987b]. It has been suggested that action potentials could be involved as signalling mechanisms in chilling injury, mechanical perturbation and invasion by pathogens [Davies et al.,1991; Davies,1987b; Wildon et al.,1989; Wildon et al.,1992], as well as for influence of light and gravity [Davies,1987a; Davies et al.,1991].

As suggested by the following experiments, endogenous electrical activities of plants may be manipulated by electric current from outside.

In the long-day plant *Spinacia oleracea* [Montavon and Greppin,1983, 1986], as well as in the short-day plant *Chenopodium rubrum* [Adamec et al.,1989; Machackova and Krekule,1991; Machackova et al.,1990], photoperiodic flower induction could be inhibited by the application of direct electric current (DC) via felt-tipped electrodes. It was concluded that DC probably interfered with the translocation of a floral stimulus from induced leaves to the apex [Adamec et al.,1989].

On the basis of our recent findings on changing rhythms in stem extension rate as markers for flower initiation, we will now consider the concept of the involvement of frequency-coded signals in the control of growth and development in plants and animals [Due,1989; Ferrier,1983]. We propose that the interaction of frequency coded (electrical) signals from the apex and the leaves interact so as to leave the rates and planes of cell division at the apex in a vegetative state or switch rhythmic mitosis to a reproductive state. Such a change of state most likely involves changes in the state of membranes, as observed in *Spinacia oleracea* [Crèvecoeur et al.,1992; Martinec et al.,1993].

Bearing in mind the circadian rhythm in transcellular current reported for *Acetabularia* [Novak and Sironval,1976], it seems possible that a circadian rhythm in bioelectricity may be of great significance in the circadian rhythmic coordination of the whole plant.

Temporal organization of development and behaviour may thus depend on a network of physiological oscillations. There are subsystems engaged in signal perception, transduction and amplification, all of these involving a sequence of primary and secondary messengers [Bowler et al.,1994; Neuhaus et al.,1993; Goldbeter and Koshland,1982]. The basic endogenous timer probably is a genetically determined circadian rhythm in the network of energy metabolism which sets the temporal order for modulation of transcription, translation and enzyme activity by environmental signals [Wagner et al.,1985; Wagner and Fukshansky,1985].

Phytochrome might manipulate membrane-bound or potentially membrane binding receptor or translocator systems. The binding to membranes could be controlled by a circadian rhythm in the competence of membrane-bound receptors or a circadian rhythm in the availability of modulator molecules like small GTP-binding proteins [Verma et al.,1994]. In developing systems, synthesis of phytochrome and receptors could be controlled by temperature dependent biochemical oscillations acting as develomental clocks [Giese and Wunderlich,1982; Satoh,1982], as the basis for so-called primary differentiation [Wagner,1976b]

Acknowledgements

We thank R. Cassada for critically reading the manuscript.

References

Adamec L, Machackova I, Krekule J, Novakova M (1989) Electric current inhibits flowering in the short-day plant *Chenopodium rubrum* L . J Plant Physiol 134:43 - 46

Aimi R, Shibasaki S (1975) Diurnal change in bioelectric potential of *Phaseolus* plant in relation to the leaf movement and light conditions . Plant Cell Physiol 16:1157 - 1162

Bertram L, Karlsen P (1994) Patterns in stem elongation rate in chrysanthemum and tomato plants in relation to irradiance and day/night temperature . Sci Hort 58:139 - 150

Bowler C, Neuhaus G, Yamagata H, Chua N-H (1994) Cyclic GMP and calcium mediate phytochrome phototransduction . Cell 77:73 - 81

Bünning E (ed) (1977) Die physiologische Uhr Springer-Verlag, Berlin

Crèvecoeur M, Crespi P, Lefort F, Greppin H (1992) Sterols and plasmalemma modification in spinach apex during transition to flowering . J Plant Physiol 139:595 - 599

Davies E (1987a) Action potentials as multifunctional signals in plants: a unifying hypothesis to explain apparently disparate wound responses . Plant Cell Environ 10:623 - 631

Davies E (1987b) Plant responses to wounding In: Davies D D (ed) The biochemistry of plants Vol 12, Academic Press, New York, pp 243 - 364

Davies E, Zawadzki T, Witters D (1991) Electrical acitvity and signal transmission in plants: how do plants know? In: Penel C, Greppin H (eds) Plant signalling, plasma membrane and change of state Imprimerie Nationale, Geneva, pp 119 - 137

De Looze SM, Wagner E (1983a) *In vitro* and *in vivo* regulation of chloroplast glyceraldehyde-3-phosphate dehydrogenase isozymes from *Chenopodium rubrum* L I Purification and properties of isozymes . Physiol Plant 57: 231 - 237

De Looze SM, Wagner E (1983b) *In vitro* and *in vivo* regulation of chloroplast glyceraldehyde-3-phosphate dehydrogenase isozymes from *Chenopodium rubrum* L II *In vitro* modulation of the isozyme pattern . Physiol Plant 57:238 - 242

De Looze SM, Wagner E (1983c) *In vitro* and *in vivo* regulation of chloroplast glyceraldehyde-3-phosphate dehydrogenase isozymes from *Chenopodium rubrum* L III The molecular basis of the aggregation phenomenon: chloroplast glyceraldehyde-3-phosphate dehydrogenase as an ambiquitous enzyme . Physiol Plant 57:243 - 249

Due G (1989) Frequency as a property of physiological signals in plants . Plant Cell Environ 12:145 -149
Ferrier J (1983) A mechanism for the regulation of ligament width based on the resonance frequency of ion concentration waves . J Theor Biol 102:477 - 486
Foyer CH, Lelandais M, Kunert KJ (1994) Photooxidative stress in plants . Physiol Plant 92:696 - 717
Frosch S, Wagner E (1973a) Endogenous rhythmicity and energy transduction II Phytochrome action and the conditioning of rhythmicity of adenylate kinase, NAD- and NADP-linked glyceraldehyde-3-phosphate dehydrogenase in *Chenopodium rubrum* by temperature and light intensity cycles during germination . Can J Bot 51:1521 - 1535
Frosch S, Wagner E (1973b) Endogenous rhythmicity and energy transduction III Time course of phytochrome action in adenylate kinase, NAD- and NADP-linked glyceraldehyde-3-phosphate dehydrogenase in *Chenopodium rubrum* . Can J Bot 51:1529 - 1535
Fukshansky L (1981) A quantitative study of timing in plant photoperiodism . J Theor Biol 93:63 - 91
Giese G, Wunderlich F (1982) *In vitro* ribosomal ribonucleoprotein transport Temperature-induced "graded unlocking" of nuclei . J Biol Chem 258:131 - 135
Goldbeter A, Koshland Jr DE (1982) Sensitivity amplification in biochemical systems . Quant Rev Biophys 15:555 - 591
Horwitz BA, Epel, BL (1978) Circadian changes in activity of the far-red form of phytochrome: Physiological and *in vivo* spectrophotometric studies . Plant Sci Lett 13;9 - 14
Jabben M, Schäfer E (1976) Rhythmic oscillations of phytochrome and its pelletability in *Cucurbita pepo* L . Nature 259:114 - 115
Kiefer S, Penel C, Greppin H (1985) Ca^{2+} and Mn^{2+} mediated binding of the glycoprotein peroxidase to membranes of *Pharbitis nil* cotyledons . Plant Sci Lett 39:37 - 43
Kiefer S, Penel C, Greppin H, Wagner E (1987) Association de peroxydases aux membranes de courgettes, de raifort et de *Pharbitis nil* . Arch Sc Genève 40:369 - 378
King RW (1975) Multiple circadian rhythms regulate photoperiodic flowering responses in *Chenopodium rubrum* . Can J Bot 53:2631 - 2638
Lucas WJ, Ding B, Van der Schoot, Ch (1993) Tansley Review No 58 Plasmodesmata and the supracellular nature of plants . New Phytol 125:435 - 476
Machackova I, Krekule J (1991) The interaction of direct electric current with endogenous rhythms of flowering in *Chenopodium rubrum* . J Plant Physiol 138: 365 - 369
Machackova I, Prospiskova M, Krekule J (1990) Further studies on the inhibitory action of direct electric current on flowering in the short-day plant *Chenopodium rubrum* L . J Plant Physiol 136:381 - 384
Marmé D (1977) Phytochrome: Membranes as possible sites of primary action . Ann Rev Plant Physiol 28:173 - 198
Martinec J, Crèvecoeur M, Crespi P, Greppin H (1993) Plasma membrane thickness and flowering induction in *Chenopodium rubrum* and in *Spinacia oleracea* . Archs Sci Genève 46:233 248
Mayer W-E, Fischer C (1994) Protoplasts from *Phaseolus occineus* L pulvinar motor cells show circadian volume oscillations . Chronobiol Int 11:156 - 164
Millet B, Botton A-M, Hayoum C, Koukkari WL (1988) An experimental analysis and comparison of three rhythms of movements in bean (*Phaseolus vulgaris* L) . Chronobiol Int 5:187 - 193
Mills JW, Lazaro J, Mandel J (1994) Cytoskeletal regulation of membrane transport events . FASEB J 8:1161 - 1165
Montavon M, Greppin H (1983) Effet sur le dévelopement de l'épinard de l'application d'un potentiel èlectrique sur le petiole d'une feuille . Saussurea (Genève) 14:79 - 85

Montavon M, Greppin H (1986) Développement apical de l'épinard et application d'un potentiel électrique de contrainte . Saussurea (Genève) 17:85 - 91
Neuhaus G, Bowler C, Kern R, Chua N-H (1993) Calcium/calmodulin-dependent and - independent phytochrome signal transduduction pathway . Cell 73:937 - 952
Novak B, Sironval C (1976) Circadian rhythms of the transcellular current in regenerating enucleated posterior stalk segments of *Acetabularia mediterranea* . Plant Sci Lett 6:273 - 283
Pahl HL, Baeuerle PA (1994) Oxygen and the control of gene expression . BioEssays 16:497-502
Paietta J (1982) Photooxydation and the evolution of circadian rhythmicity . J Theor Biol 97:77 - 82
Pickard BG (1994) Contemplating the plasmalemmal control center model . Protoplasma 182:1 - 9
Quail PH (1975a) Particle-bound phytochrome: association with a ribonucleoprotein fraction from *Cucurbita pepo* L . Planta 123:223 - 234
Quail PH (1975b) Particle-bound phytochrome: The nature of the interaction between pigment and particulate fractions . Planta 123:235 - 246
Ruiz-Fernandez S, Wagner E (1994) A new method of measurement and analysis of the stem extension growth rate to demonstrate complete synchronization of *Chenopodium rubrum* plants by environmental conditions . J Plant Physiol 144:362 - 369
Ruiz-Fernandez S, Wagner E (1989) Flowering in *Chenopodium rubrum* Light control of stem elongation rate (SER) as a systemic marker for flower induction . Flowering Newsletters 8:14 - 19
Sandelius AS, Penel C, Auderset G, Grightman A, Millard M, Morré DJ (1986) Isolation of highly purified fractions of plasma membrane and tonoplast from the same homogenate of plant cells by free flow electrophoresis . Plant Physiol 81:177 - 185
Satoh N (1982) Timing mechanisms in early embryonic development . Differentiation 22:156
Srere PA, Mosbach K (1974) Metabolic compartmentation: symbiotic, organellar, multienzymic and microenvironmental . Ann Rev Microbiol 28:61 - 83
Tanada T (1983) Photogeneration of a bioelectric field by red and far-red irradiation in soybean hypocotyls . Plant Cell Environ 6:69 - 72
Verma DPS, Cheon Ch-III, Hong Z (1994) Small GTP-binding proteins and membrane biogenesis in plants . Plant Physiol 106:1 - 6
Vince-Prue D, Gressel J (1985) *Pharbitis nil* . Japanese morning glory In: Halevy AH (ed) CRC Handbook of flowering, Vol IV CRC Press Inc Boca Raton, Florida, pp 47-81
Wagner E (1976a) Endogeneous rhythmicity in energy metabolism: Basis for timer-photoreceptor-interactions in photoperiodic control . In: Hastings JW, Schweiger HG (eds) Dahlem Konferenzen, Abakon Verlagsgesellschaft, Berlin, pp 215 - 238
Wagner E (1976b) Kinetics in metabolic control of time measurement in photoperiodism . J Interdiscipl Cycle Res 7:313 - 332
Wagner E (1976c) The nature of photoperiodic time measurement: energy transduction and phytochrome action in seedlings of *Chenopodium rubrum* In: Smith H (ed) Light and plant development . Proceedings of the 22 nd Nottingham Easter School in Agricultural Sciences, Butterworth, London
Wagner E (1977) Molecular basis of physiological rhythms . In: Jennings DH (ed) Integration of activity in the higher plant, Society for Experimental Biology, Symposium 31, Cambridge University Press, pp 33 - 72

Wagner E (1983) Circadian rhythms: The basis for information processing in eukaryotes during adaptation to seasonal changes in photo- and thermoperiods . In: Montagnoli G, Erlanger BF (eds) Molecular models of photoresponsiveness . Plenum Press, New York, pp 197 - 202

Wagner E, Bonzon M, Greppin H (1985) Membrane-oscillator hypothesis of metabolic control in photoperiodic time measurement and the temporal organization of development and behaviour in plants . In: Packer L (ed) New developments and methods in membrane research and biological energy transduction, Plenum Pub Co , pp 525 - 546

Wagner E, Cumming BG (1970) Betacyanin accumulation, chlorophyll content and flower initiation in *Chenopodium rubrum* as related to endogenous rhythmicity and phytochrome action . Can J Bot 48:1 - 18

Wagner E, Fukshansky L (1985) Die zeitliche Organization des eukaryotischen Energiestoffwechsels als Grundlage für die Signalverarbeitung im Photo- und Thermoperiodismus . Ber Dtsch Bot Ges 98:35 - 52

Wagner E, Haertle U, Kossmann I, Frosch S (1983) Metabolic and developmental adaption of eukaryotic cells as related to endogenous and exogenous control of translocators between subcellar compartments . In: Schenk H, Schwemmler W (eds) Endocytobiology II, Walter de Gruyter & Co , Berlin, New York, pp 341 - 351

Wagner E, Ruiz-Fernandez S, Normann J, Bonzon M, Greppin H (1993) Chronobiology - Spatio-temporal organization of living systems - The induction of flowering . In: Greppin H, Bonzon M, Degli Agosti R (eds) Some physico-chemical and mathematical tools for understanding living systems . University of Geneva, pp 109 - 124

Walter P, Blobel G (1981) Translocation of proteins across the endoplasmic reticulum III Signal recognition protein (SRP) causes signal sequence-dependent and site-specific arrest of chain elongation that is released by microsomal membranes . J Cell Biol 91:557 - 561

Wildon DC, Doherty HM, Eagles G, Bowles DJ, Thain JF (1989) Systemic responses arising from localized heat stimuli in tomato plants . Ann Bot 64:691 - 695

Wildon DC, Thain JF, Minchin PEH, Gubb IR, Reilly AJ, Skipper YD, Doherty HM, O'Donnell PJ, Bowles DJ (1992) Electrical signalling and systemic proteinase inhibitor induction in the wounded plant . Nature 360:62 -65

Yamaguti Y (1932) Über elektrische Potentialveränderungen an periodisch sich bewegenden Primärblättern von *Canavalia ensiformis*, D C . Bot Mag (Tokyo) 46:216 - 222

Yang S, Xiang G, Zhang S, Lou C (1992) Electrical resistance as a measure of graft union . J Plant Physiol 141: 98 - 104

11 The Plasma Membrane of Acetabularia: an Integrating Function Regulated by Circadian Rhythmicity

T. Vanden Driessche[1], J.-L. Guisset[2], G.M. Petiau-de Vries[3] and T. Gaspar[4].
[1]Dept. de Biologie moléculaire. Université Libre de Bruxelles. 46, av. des Ortolans, B-1170 Bruxelles. Belgium.
[2]Pool de Physique. Université Libre de Bruxelles. Campus de la Plaine. CP 222. Bd du Triomphe. B-1050 Bruxelles.
[3]Chimie générale I. Fac. Médecine. Université Libre de Bruxelles. Campus du Solbosh CP. 160/18. 50 av. FD Roosevelt. B-1050 Bruxelles.
[4]Hormonologie fondamentale et appliquée. Institut de Botanique B 22. Sart Tilman. B-4000 Liège. Belgium.

Summary

Numerous receptors and structures in the plasma membrane allow the cell to sense environmental signals as well as internal status and to adjust its own activities. We review the circadian rhythms affecting the Acetabularia PM, both from direct and indirect experimental evidence . We consider the CR in bioelectric properties, fatty acid composition and glycosylation degree. We indicate that changes in the phosphorylation degree might occur. The photosensitivity of growth and cap morphogenesis is modulated by a CR, as well

Abbreviations: ABA, abscisic acid; CaM, calmodulin; CDPK, Ca2+-dependent phosphokinase; CR, circadian rhythm; CT, circadian time (time 0 is the transition time from dark to light period); chlpl envel, chloroplast envelope; CW, cell wall; DD, constant dark; DAG, diacylglycerol; 6-DMAP, 6-dimethylaminopurine; ER, endoplasmic reticulum; IAA, indole-3-acetic acid; IP, inositol 4-monophosphate; IP2, inositol 1,4-bisphosphate; IP3, inositol 1,4,5-trisphosphate; IPS3, triphosphothioate; LD, light-dark conditions; LL, constant light; LOX, lipoxygenase; PIP2 phosphatidyl inositol 4,5-bisphosphate; PK, phosphokinase; PLA2, phospholipase A2; PLC phospholipase C; PM, plasma membrane; R, receptor; SHAM, salicylhydroxamic acid; ser/thr, serine/threonine symp, symport; Th-S2, thioredoxin in its reduced form; thylak, thylakoid; tr, transporter; tyr ki, tyrosine-protein kinase; Vac, vacuole; X-Gal, 5-bromo-4-chloro3-indolyl-beta-D galactopyranoside.Note: the photosynthetic enzymes, in the chloroplast of fig. 1, are not given in full here.

as (delayed) morphogenesis after a SHAM treatment. The transducing chain, activated by the photoreceptor is also oscillating (the IP3, Ca^{2+}-calmodulin pathway). An NADP(H) oxido-reductase is located in the PM and presumably related to the redox state of the non chloroplastic pyridine nucleotides. The cytoplasmic thioredoxin h, an important redox regulator has been isolated in Acetabularia, but its low amount prevents analyses along the 24 h cycle. Finally, the system peroxidases/ascorbic acid was shown to be present, but, for similar reasons, it has not been analyzed during the daily cycle. Three phytohormones have been studied: IAA, ABA and ethylene. All three induce time-dependent morphogenetic responses. The metabolism of ethylene itself is highly regulated by a CR. The numerous responses from the membrane act in a concerted manner, owing to the integrating function of the PM and the interrelations of the cellular pathways. We propose that, in addition, circadian rhythmicity imposes a regulatory control on the activities at the level of the PM.

11.1 Introduction

The plant plasma membrane not only limits the cytoplasm but, being the interface between cell and environment has been organized, in the course of evolution, in an organelle endowed with receptors and the task of sensing external signals. Its activity is influenced by the physical state of the cytoplasm on the one hand and by that of the cell wall and apoplast on the other.
From the cytoplasm a number of developmental, hormonal signals are directed to the plasma membrane; the redox state of the cytoplasm and its concentrations in a variety of molecules and ions - Ca^{2+} concentration in particular - exert a control on the plasma membrane activity. The cell wall may retain or "electrophoreze" a number of molecules from the so-called periplasm (or apoplast), the fluid stationary phase contained between the plasma membrane or plasmalemma and the cell wall: it is outside the cytoplasm and penetrates the CW (Grignon and Sentenac, 1991). Thus the plasma membrane (fig. 1.) is by no means an independent organelle, it cannot be dissociated from its cellular surroundings. As a consequence, it is not constant.

In the plasma membrane are located a collection of receptors, pumps and channels. The presence of some of them has been demonstrated in the particular case of Acetabularia. The presence of others is infered from indirect evidence

Fig.11.1. Schematic representation of the Acetabularia plasma membrane. Some structures appear to be specific, such as the Cl^- pump, and, possibly the insulin receptor, but the scheme applies more generaly, to typical plant cell PM. Some receptors, channels etc...are deduced from comparison. References not given in the text for transporters are from Blatt and Thiel (1993) and Sussman (1994), for tyrosine protein kinase receptor, from van der Geer et al. (1994), for serine/threonine kinase, from Schulman (1991) and Sussman (1994).

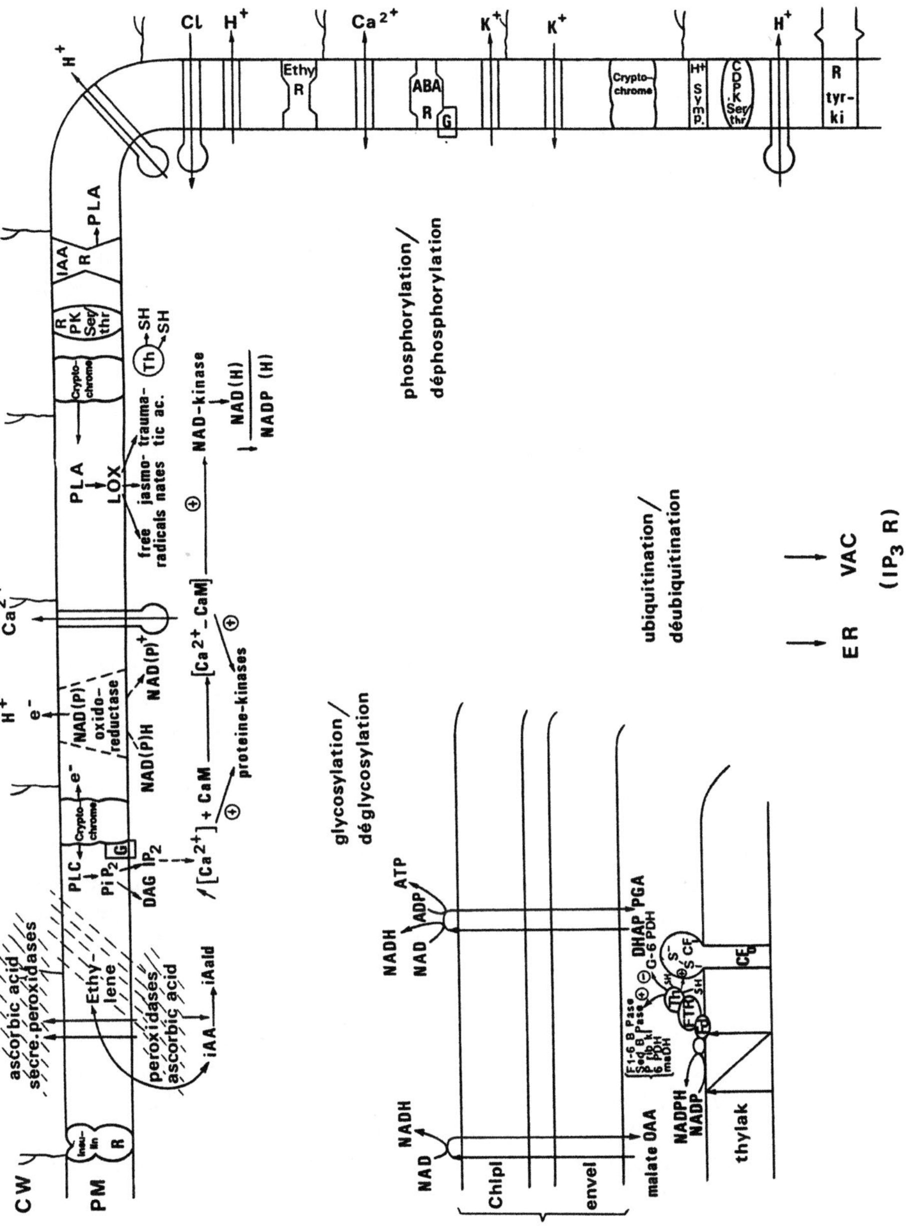
CW
PM
insulin R
ascorbic acid
sécre.peroxidases
Ethylene
peroxidases
ascorbic acid
iAA
iAald
PLC
PiP2
DAG
IP2
G
Cryptochrome
e-
$[Ca^{2+}]$ + CaM
proteine-kinases
NAD(P)H
NAD(P) oxido-reductase
$NAD(P)^+$
H+
Ca^{2+}
$[Ca^{2+}\_CaM]$
PLA
LOX
free radicals
jasmonates
traumatic ac.
NAD-kinase
NAD(H)
NADP (H)
Th
SH
R PK Ser thr
IAA R
Cl
Ethy R
ABA R
K+
Cryptochrome
H+ symp.
CDPK Ser thr
R tyr-ki
phosphorylation/
déphosphorylation
ubiquitination/
déubiquitination
glycosylation/
déglycosylation
ER
VAC
(IP_3 R)
NADH
NAD
ATP
ADP
PGA
DHAP
G-6 PDH
malate OAA
NADPH
NADP
Chlpl
envel
thylak
CF

and analogy with other plant systems. Since the aim of the chapter is to provide an overview on the control the plasma membrane exerts on the cell's rhythmic activity, they have to be taken into account.However, the most obvious evidence of plasma membrane circadian changes will first be reported.

Then, we shall deal with the location of the photoreceptor and the IP3 transduction pathway. We discuss,the evidence for other signalling transducing systems, in particular, the redox one.

We then examine phytohormone perception and activity as well as location of its synthesis in the case of ethylene.We conclude with the intricacy steming from plasmalemma diverse receptors and the pathways they activate. Finally we show how this view fits in our CR model adding a new control.

11.2 Basic Rhythmic Activity of the Plasma Membrane

11.2.1 Electrical Properties and Pumps

Acetabularia displays a membrane potential of -170 mV resulting mostly from the Cl^- electrogenic pump localized in the plasma membrane which actively pumps in Cl^- ions. They do not intervene alone, however (Gradmann, 1975). The resting potential is also affected by K^+ (but no other ion). K^+ diffusion is complex (Mummert and Gradman, 1976) and, in addition, K^+ ions are lost through action potentials (Gradmann, 1976). A scheme of some of the major transport mechanisms in Acetabularia are given in fig. 1.

Noteworthy is the change in bioelectric characteristics in the course of development.

In regenerating algae, longitudinal growth is correlated with an apico-basal bioelectric potential gradient and with the appearence of recurrent action potentials propagating from apex to rhizoid. These bioelectrical characteristics are not perturbed by Ca^{2+} ions, Mg^{2+}ions or by the A-23187, the Ca^{2+}, Mg^{2+} ionophore (Goodwin and Pateromichelakis, 1979).

The transmembranar potential varies from tip to base and gives rise to a transcellular current of critical importance for Acetabularia development: voltage clamping to zero of the transcellular current inhibits regeneration (Novak and Bentrup, 1972). The transcellular current itself displays a circadian rhythm (CR) whose amplitude is modulated by light intensity.

The CR is superimposed with high frequency oscillations whose distribution in time is also circadian (Novak and Sironval, 1976).

11.2.2 Does the Fatty Acid Composition of the PM Undergoes Periodic Changes?

There is no direct measurements of plasma membrane components of Acetabularia. However, it is generally assumed that a very large part of the cell's lipids are located in the plasma membrane. The fatty acid composition of

Acetabularia (whole algae) has been examined in both light-dark (LD) and constant light (LL) (Jerebzoff and Vanden Driessche, 1983). In whole algae, there is a sharp peak in total fatty acids at about 9 h CT due to accumulation of oleic acid. Saturated fatty acids in C 16 and unsaturated fatty acids in C 18 display a minor peak at 6 h.Since presumably the non-chloroplast fatty acids are mainly membranar -and more specifically located in the plasma membrane- we surmise that the fatty acids plasma membrane composition is subjected to circadian modulation: the organelle itself presumably changes with time in the 24 h cycle.Among possible effects of circadian changes in lipid composition of the PM is the modulation of plasma membrane-bound cation-pumping ATPases (Cooke and Burden, 1990) besides changes in physical properties.

11.2.3 Is There a CR in Plasma Membrane Glycosylation?

Glycoproteins are important components of the plasma membrane, mediating number of cell-cell recognition processes. They reside prominently in the PM but also on various internal membranes. In addition, they can occur unlinked with membranes. Our studies with inhibitors of glycosylation revealed how important glycosylation is in Acetabularia development: tunicamycin, an inhibitor of N-linked glycoprotein synthesis, given in pulses of 3 or 4 hours (10 μg ml/l) two successive days always inhibited growth but more severely during the light period. One pulse may inhibit growth during the light period and stimulate it during the dark one (it may also have little effect during both periods). Cap formation is inhibited between time 0 and 7 h. During the dark or subjective dark period, it is often stimulated or not affected. Cap formation is also inhibited in anucleate algae treated during the light period. Ultimately, however, development is completed.

To approach more directly the intracellular location of at least one facet of the glycoprotein metabolism, we carried out a cytochemical study of β-galactosidase, one of the enzymes involved in its metabolism. We used a substrate analogue (X-gal). The enzyme is present in both the cytoplasm and the cell wall. The β-galactosidase activity was then exhaustively studied in the whole alga and quantified by fluorescence spectrometry. One of the results is the demonstration of a well-delineated biphasic CR in activity (Vanden Driessche et al., 1992).

In addition, many hormone receptors appear to be glycosylated in a circadian time dependent process (see Petiau-de Vries, chap.3): this adds to the presumption that at least part of the glycosylation is concerned with the PM. Circadian modulation of membrane glycosylation is a likely hypothesis. A mutant lacking N-acetyl glucosaminyl transferase I has been isolated (von Schwaeven et al., 1993) by screening a population of Arabidopsis mutants with a serum directed against a β 1-2 xylose. It is unable to synthetize Golgi-modified complex N-linked glycans. It is able, however, to complete normally development.

Thus, in both Acetabularia and Arabidopsis, glycosylation, important as it is, is not necessarily critical for development.

11.2.4 Phosphorylation of PM Molecules?

Phosphorylation/dephosphorylation plays a very important role in plant cells as well as in any cell. It is possibly involved in CR operation.

Reversible phosphorylation/ dephosphorylation concerns a number of proteins, enzymes and PM associated receptors (tyrosine-linked receptors) and channels. Siegelbaum reviews (1994) the consequences of ionic channels phosphorylation in electrical properties of some animal model systems. The existence of tyrosine receptor kinases in plants has been questioned because kinase receptors cloned from plants display closest sequence homology with serine/threonine kinases. However, their distribution among Eukaryotes is such (van der Geer et al., 1994) that their existence appears to be likely, in addition to the of serine /threonine receptor kinases. Better known is the Ca^{2+}-dependent protein kinase which plays a key role in regulating the activity of the H^+ pump, in protoplasmic streaming (a CR-modulated process) and in controling cytoskeleton-plasma membrane association.

The question has not been examined in Acetabularia.

Finally, we would like to emphasize, after Shenolikar (1994) that, on the one hand, activated protein kinases lead to a cascade of events and on the other hand, protein phosphatases reverse their action, acting as negative regulators. The former are responsible for signal transduction whereas the latter may act constitutively.

Several authors have implicated the phosphorylation/dephosphorylation process in the CR operation.Limiting ourselves to Gonyaulax, another unicellular Protoctist, circadian changes have been demonstrated in both protein synthesis and phosphorylation degree (Schröder-Lorenz and Rensing, 1987).

Recently, blocking the circadian oscillator was obtained by a treatment with 6-dimethylaminopurine (6-DMAP), an inhibitor of protein kinase. Gonyaulax cultures, continuously exposed to 6-DMAP, display a period lengthening in a concentration-dependent manner. Shorter treatments at high concentration apparently stop the "oscillator" and pulses induce delays at all times of the circadian cycle but of different importance depending on the CT. In addition, 6-DMAP prevent light-induced advances in vivo (Comolli et al., 1994).

PM proteins phosphorylation is considered in the next section as well as the protein kinase as influenced by blue light.

11.3 The Plasma Membrane Photoreceptor.

11.3.1 The Cryptochrome and the SHAM-Sensitive Associated Molecule.

The single photoreceptor of Acetabularia deserves a special section. It will be studied together with the salicylhydroxamic (SHAM) acid sensitive molecule with which it is closely associated.

Blue light (maximum wavelength efficiency: 415 nm) is the only but stringent light requirement for both growth and differentiation in Acetabularia. Cap morphogenesis requires a higher light intensity than growth (Clauss, 1977. Schmid, 1984).

The blue light receptor has been identified as the cryptochrome, a flavin-cytochrome b complex (Caubergs et al., 1984). It is responsible for the typical light-induced absorbance change (LIAC). This response, in higher plant plasma membranes is shut off by salicylhydroxamic acid, quenching the excitation of the cryptochrome. In Acetabularia, a direct extinction of LIAC by SHAM could not be detected (Caubergs et al., 1984). However, it appears that the SHAM-sensitive molecule acts close to the cryptochrome within the plasma membrane since it causes disruption of the PM: SHAM was allowed into the cell by severing the stalk into SHAM containing sea-water. The medium was replaced half an hour later by a medium without SHAM and the fragments allowed to heal. Cytoplasts, droplets of cytoplasm, were released and, at the same time, the cytoplasm retracts and the new plasma membrane is formed several cms below the section. Its relatedness with the photoreceptor derives from physiological effects: it alters the PM and severely delays regeneration. These observations show that the SHAM-sensitive molecule pertain to the plasma membrane, as does the receptor (Vanden Driessche and Caubergs, 1985). The possibility that SHAM effects could be due to peroxidase activity (Rubinstein and Luster, 1993) is not likely since KI and phenyl acetic acid exert similar effects on light-induced absorbance changes or on a closely related process ascertained by their action on phototropism and geotropism in corn coleoptile. The time-dependency of the light effects is comparable (see below).

Acetabularia development -essentially growth and cap morphogenesis- is modulated by a CR. Light and dark perturbations induce different delays in cap formation, suggesting that light shown in the early dark period induces the greatest delay (Vanden Driessche, 1984).

SHAM, KI and phenyl acetic acid sensitivities in the subsequent period of time are reflected by transient effects on morphogenesis. Depending on the nature of the inhibitor and the concentration used, delays or both accelerations and delays occur, but always the effects are more accentuated in the dark period than in the light one (Vanden Driessche and Caubergs, 1985).

Recently, Briggs and his collaborators demonstrated that blue light activates a protein kinase (Reymond et al., 1992). The same group reported blue-light-induced phosphorylation of a plasma membrane polypeptide of about 120 kD in Pea (Short et al.,1993). The photoreceptor, the kinase (at least a kinase moiety) and the substrate are localized in the PM fraction (Short et al.,1993). This specific kinase could have a wide distribution. Interestingly, in Arabidopsis, mutant studies showed a correlation between the phosphorylation level and the first phototropic response (Reymond et al.,1992).

11.3.2 The Receptor for Light Signals Synchronizing the Chloroplast CR.

Remarkably, the various CR displayed by the chloroplasts, the CR in photosynthesis being the most prominent one, are synchronized by the plasmalemma located cryptochrome (as reviewed by Vanden Driessche, 1994).

11.3.3 The Transduction Chain Originating from The Photoreceptor. Postulated CR of the PM Itself.

As indicated above, Acetabularia possesses a single photoreceptor, the cryptochrome. The first steps in transduction take place, necessarily, in the PM. There is evidence in Acetabularia for some molecules of the classical chain conveying external signals into cells, of plant as well as of animal systems: the protein G-mediated phospholipase C activation. Activated, PLC cleaves the membrane lipid phosphatidylinositol 4,5-bisphosphate (PIP2) in two products, inositol 1,4,5-trisphosphate (IP3) and diacylglycerol (DAG). IP3 has been identified in Acetabularia (Chaignon et al., 1993a) as well as other phosphoinositides, inositol 4-phosphate (IP) and inositol 1,4-bisphosphate (IP2) (Chaignon et al. 1993b). Its physiological role has been assessed by the effects of triphosphothioate, IPS3, an IP3 analogue (Vanden Driessche et al., 1993) and by the analysis of the effects of lindane (hexachlorocyclohexane), an inhibitor of phospholipids turnover (Chaignon and Vanden Driessche, 1993).

This part of the transducing chain is contained within the plasma membrane. The further steps are cytoplasmic: we know that both calcium (see Vanden Driessche, 1990 for a review) and calmodulin (Cotton and Vanden Driessche,1987) are involved in signaling in Acetabularia. Thus a few steps of this complex pathway (Berridge and Irvine, 1989, Berridge,1993) were formally identified in the alga. In addition, we observed that pulses of specific inhibitors of any step of the chain, SHAM, lindane, IPS3, lanthanum, a competitive inhibitor of Ca^{2+} , as well as light shown at the "wrong" time (light during the dark period or darkness during the light one) exert maximum effect during the same part of the circadian cycle, i.e., in this case, in the first part of the dark or subjective dark period (as already proposed in 1985, Vanden Driessche et al.).

This particular time-dependent sensitivity, involving several molecules of the PM is suggestive of a circadian oscillation displayed by the PM itself. Could these various molecules whose sensitivity changes with time in the 24 h cycle be part of various oscillators or does the PM oscillation imposes a CR to the whole transducing chain or chains?

We wish to underline that another (at least) alternative transducing chain may be activated by the cryptochrome: possibly via a serine/threonine kinase or via a phospholipase A2. We do not know how thioredoxin is reduced by light as indicated by recent experiments (see below).

11.4 The Plasma Membrane NADP(H)-Oxido-Reductase.

This plant plasma membrane enzyme may be endowed with often under-estimated regulatory functions. Recently, however, plasma membranes molecules capable of regulating redox processes were reviewed by Moller and Crane (1990) and by Rubinstein and Luster (1993). Asard and Caubergs also examine the problem (chap.7).

Acetabularia does possess a typical oxido-reductase enzyme in its plasma membrane (Asard. 1991). The enzyme is entirely comparable to that studied in other plants (Horemans et al.,1993). No doubt that it influences the cytoplasmic redox equilibrium since pulses or treatments with exogenously applied NAD^+ (Vanden Driessche, 1972) or $NADP^+$ (Glory and Vanden Driessche, 1985) exert physiological effects.

Acetabularia cytoplasm is equipped with the machinery that regulates the redox potential which presumably exert a control on development: the non-chloroplastic pyridine nucleotides and thioredoxin.

11.4.1 The Non-Chloroplastic Pyridine Nucleotides

We evidenced both developmental and circadian changes. NAD^+ and $NADP^+$ concentrations display CR in phase according to a biphasic mode. For their reduced counterparts, no CR could be ascertained because of their low concentration and strong fluctuations (Van Langendonckt et al. 1991).

11.4.2 Cytoplasmic thioredoxin

Thioredoxin (in the h form) is also present in the alga. Specifically we identified, in addition to the chloroplast thioredoxins, an NADP/thioredoxin system in the extrachloroplastic fraction (Van Langendonckt and Vanden Driessche, 1992a).

In order to localize thioredoxin in the alga (in vivo), we used the 'Thiolyte' monobromobimane, a highly specific thiol-labeling agent, described by Kosover and Kosover (1947). Fluorescence was localized only on the external part of the cell, close -at microsopic resolution level- of the plasma membrane (Van Langendonckt and Vanden Driessche, 1992b).

Again, the concentrations were too low to allow a rhythm study with the available, although already sensitive methods. We demonstrated however that thioredoxin distribution changes with development (Van Langendonckt and Vanden Driessche, 1992b) and that its amount is dependent on blue light (Van Langendonckt and Vanden Driessche, 1994).

However, we were unable to induce in vitro changes in thioredoxin after illuminating the cryptochrome with blue light (Caubergs, Van Langendonckt, Asard and Vanden Driessche, unpublished).

11.5 The Peroxidase/Ascorbic Acid System

Peroxidases and, more specifically the system peroxidases/ascorbic acid, are integral parts of the plasma membrane (Penel and Castillo, 1991) . Such an ascorbate peroxidase/ascorbic acid system was identified in Acetabularia and shown to be operative in the periplasm (Castillo et al. 1986). Peroxidases are distributed along the highly polarized alga according to a negative apico-basal gradient, reiterated along hair whorls (Vanden Driessche et al., 1984). The peroxidase/ascorbic acid system is developmentaly regulated: changes in the levels of the enzyme and its substrate in the cytoplasm and periplasm were demonstrated using biochemical and cytochemical procedures. It is also greatly influenced by light conditions. Comparing Acetabularia cells grown in good conditions (1200 lux) and cells grown under low light intensity (700 lux), it was found that peroxidase activity is twice as high in Acetabularia cultivated at low light intensity than in normal light. Moreover, peroxidase distribution is not the same under these two conditions: more peroxidases are found in the periplasm when conditions are unfavourable.

Ascorbic acid undergoes parallel changes.

Quite recently, it has been demonstrated that a transport system for ascorbate is present in the plant plasma membrane; in vitro experiments showed that it is a carrier-mediated process (Horemans et al. , 1994)

Catalase was found to be present at very low level but is unlikely to play a role in H_2O_2 catabolism. In addition, it is inhibited by physiological concentrations of ascorbic acid (Castillo et al.,1986).

The possible existence of a CR in peroxidase or ascorbic acid was not studied.

General Conclusion of the Sections 11.3 to 11.5 It appears that at least two transducing chain operate in Acetabularia. First, the classical receptor-G protein-IP3-Ca^{2+}-calmodulin and second, a pathway potentially conveying redox signals. Acetabularia contains the elements required in order to regulate the redox potential.

The most interesting influence of blue light on thioredoxin raises the question of whether the cryptochrome acts in a more or less direct manner on thioredoxin or if it does so by influencing other molecules in the plasma membrane.

11.6 The Phytohormone Receptors

All five major phytohormones are presumably present in Acetabularia: auxin, i.e. indole acetic acid (IAA), abscisic acid (ABA) and ethylene have been identified (respectively by Vanden Driessche and Delegher, 1977, Kof and Kefeli, 1979, Vanden Driessche et al, 1988). Gibberellic acid and cytokinins are likely to be present as well since they exert effects on morphogenesis when exogeneously applied (preliminary evidence, Vanden Driessche, 1985).

All three phytohormones, IAA, ABA and ethylene exogenously applied affect the developmental program as well as IAA antagonist, morphactin. The major effect is a time-dependent delay or acceleration in cap formation.

The effects exerted by phytohormones presumably imply plasma membrane receptors.

In addition, a phase shift of a chloroplast CR - photosynthesis- is observed in algae treated with anyone of IAA (Vanden Driessche, 1979), ABA (Vanden Driessche et al., 1993) and ethylene (Vanden Driessche et al., 1994).

11.6.1 The IAA Receptor. Mechanism of one Possible Function of IAA in Acetabularia

Our early attempts to study the IAA binding-protein in Acetabularia in the beginning of the 80ies were unsuccessful because of internalization problems (as can happen with PM receptors). In other plant systems, however, much evidence was gained in favour of the existence of IAA receptors, at least one of them residing in the PM, and thought to function in an autocrine manner (Klämt et al. 1990, Napier and Venis, 1991, Palme and Schell, 1991). The receptor appears to be exposed at the exterior face of the PM (Venis et al. 1990) and to act through a G-protein (Klämt, 1990).

However, the notion of receptor in the case of IAA has been questioned (Jones, 1994).

The effects of IAA, on Acetabularia development are circadian time-dependent (Vanden Driessche, 1984). This possibly originates from changes in the PM but does not rule out the possibility that it originates from oscillating cytoplasmic transducing pathways. IAA phase shifs the CR in photosynthesis (Vanden Driessche, 1979).

The mechanism by which one of the IAA functions operates is especially interesting. Studies based on gene analysis of Arabidopsis wild type and auxin-resistant mutant, AXR1) suggest that it encodes a protein which shares sequence identity with the E1 enzyme of the ubiquitin activation pathway (Ottoline et al.,1993). This putative function appears relevant in the case of Acetabularia, since our experiments have shown that IAA acts at a post-transcriptional level in controling cap formation (Vanden Driessche, 1984). It does not exclude the possibility of transcriptional effects on whole cells. In Zea mays for example, a gene coding for a presumptive receptor was identified, Zmpk1, which shares structural motifs with the raf family of serine-threonine kinase (see Palme and Schell, 1991). Moreover, nuclear IAA receptors appear to be present in different systems.

Since anucleate Acetabularia are able to carry on morphogenesis, and since they are responsive to phytohormones, post-transcriptional processes induced by phytohormoes deserve special attention. Ubiquitination is not the only post-transcriptional action of IAA. Effects of auxin on ion transport across the PM have been reported: it promotes H^+ extrusion probably by controling the H^+-ATPase through a series of steps, the first one involving an A-type phospholipase (Blatt and Thiel, 1993). Scherer and colleagues established that auxin stimulates a phospholipase A2 (Scherer and André, 1989) and that there is a phospholipid-stimulated protein kinase in plants (Martiny-Baron and Scherer (1989). However, IAA also increases the phospholipds turnover and, presumably cytoplasmic Ca^{2+} concentration (see section 3.3).

11.6.2 The ABA Receptor. Mechanism of One Possible Function of ABA in Acetabularia

In the case of Acetabularia, a treatment with exogenously applied ABA exerts time-dependent effects on morphogenesis and on the photosynthesis CR phase (Vanden Driessche et al., 1993). Anucleate algae are sensitive to the hormone action.

Strong evidence for an ABA receptor was given by different authors in several systems in the eighties. Similar results were obtained by immunological assays (as, for ex., Wan and Hasenstein, 1993). At the PM level, ABA presumably acts through a G-protein.

Again, Arabidopsis mutants allowed to select ABA defective clones (the mutation severly impairs different ABA responses), to compare their DNA with that of the wild type and to identify the gene sequence (Leung et al. 1994, Meyer et al., 1994). The ABAI1 gene contains, at its amino-terminal end, the characteristic structure known as an EF-hand calcium binding site suggesting that the protein activity is modulated by Ca^{2+}. The carboxy-terminal domain is related to a serine/threonine phosphatase 2C. This combines a calcium-binding site and a phosphatase activity in the ABA gene product: it could couple ABA and Ca^{2+} signals with phosphorylation-dependent pathways. It constitutes a way in which the phytohormone might act post-transcriptionally as we observe in Acetabularia.

It does not exclude other mechanisms, including transcriptional effects.

Indeed, other ABA mutants revealed the existence of genes involved in Arabidopsis developmental pattern (see Goldberg et al.,1994). They may be involved in Acetabularia growth as well as in cap-development steps as suggested by the changes in sensitivity towards ABA (Verhoustraeten, 1987).

ABA modulates ion fluxes through 4 classes of ion channels: K^+, Ca^{2+}, anions (Blatt and Thiel, 1993).

11.6.3 Ethylene Synthesis, Receptor and Function

As the two other phytohormones, exogeneously applied ethylene induces physiological responses.

We have evidenced its presence in the cell as well as in the periplasm (Vanden Driessche et al., 1988).

The hormone can be synthetized in Acetabularia by the plasma membrane, as shown by experiments with microsomal fractions(Kevers et al., 1986).

The production of ethylene is modulated by a CR: the ethylene synthetizing activity of the PM is periodic.

The production of ethane is similarly rhythmic but could be linked to experimental procedures: ethane is generally considered as a catabolic product of unsaturated lipid hydroperoxides, as indicated by Leshem et al., (1992). Ethylene production varies according to the algal developmental stage. The production rate is decreasing with progression in the developmental cycle but, in contrast, the capacity of algae to convert the ethylene precursor, ACC (1-aminocyclopropanecarboxylic acid) varies in inverse manner in the same algae: while ethylene production is decreasing, that of ethane is increasing (ACC acting possibly as free radical scavenger (Kevers et al. 1989). Later in development, when caps are on the brink of desintegration, the mechanism for regulating the phosphatidyl-linoleyl(-enyl) cascade (Leshem 1987) appears possible in the alga.

In the case of this hormone, elegant molecular work going from ethylene insensitive mutants to the ethylene receptors and to the coding genes, allowed the cloning of the Constitutive Triple Response mutant gene 1. It encodes a 92 kD putative serine/threonine protein kinase. Its sequence is related to the Raf protein kinase family -typical signal-integrating enzymes, linking the PM and the cell's interior. By comparison, the product of the gene, EIN1, could be a kinase active in the absence of ethylene, presumably phosphorylated and acting on the gene product of another ethylene gene, EIN3, which could be a transcription factor (summarized by Theologis, 1993).

In anucleate Acetabularia, the signal transducing chain could operate at post-transcriptional level whereas both transcriptional and post-transcriptional effects could be carried out in whole cells.

11.6.4 Other Phytohormones are Likely to be Present in Acetabularia. Some Could be Acting at the Level of the PM

Presumably some other phytohormones, either known as "major" ones or more recently evidenced ones might also involve the PM for binding.

11.7 The Insulin Receptor. Its Role in Acetabularia

This receptor has been identified and extensively studied by Legros and colleagues (Henry et al., 1979). It is probably glycosylated (Legros et al.,1975). It allows the alga to respond to some external signals-and to modify polysaccharide metabolism in regulating the number of receptors, as well as chloroplast polysaccharide content. The responses are nucleus-dependent (Legros et al. 1973b). It could possibly belong to the insulin receptor subfamily of the protein tyrosine kinases (van der Geer et al., 1994). Most of the receptors protein tyrosine kinases are glycosylated. Indeed, Acetabularia is subjected to glycosylation as discussed above. The receptor dimerizes and elicits complex signaling and metabolic responses. Recently, Pause and colleagues (1994) demonstrated that human insulin stimulates protein synthesis in phosphorylating the translation initiation factor eIF-4E: hyperphosphorylation of the factor causes its dissociation from its complex with eIF-4F thereby relieving translation inhibition.

In Acetabularia, the actual ligand is unknown but is hypothetized to be an insulin-like molecule and to act as an autocrine hormone. Since the non-chloroplast polysaccharide metabolism of Acetabularia is modulated by CR (preliminary experiments in Vanden Driessche et al., 1993), it should be interesting to examine if the insulin membranar receptor changes periodically.

11.8 Discussion and Conclusion

The various activities of the plasma membrane of Acetabularia demonstrate that the organelle is modulated by circadian rhythmicity: its bioelectrical activity, its variable sensitivity to phytohormones, circadian changes in ethylene synthesis, sensitivity to inhibitors. A number of membrane molecules influence others. For example, lipids modulate cation-pumping ATPases (Cooke and Burden, 1990) and there is a protein kinase which is modulated by phospholipids. Phytohormones affect the membrane. ABA brings about lipid phase alterations via a probable lipid-protein lattice modification (Puroit et al.,1992). Another example, involving the same phytohormone is the activation of a Ca^{2+} channel by ABA and resultant inhibition of a K^+ channel (Ding and Pickard, 1991). A number of ion transporting systems exert effects on others as has been discussed by Schroeder and Hedrich (1989). It is likely that the light signal percieved by the photoreceptor activates not only the specific transduction pathway (via PLC, IP3 etc..., but other pathways as well: the redox signalling is probably one of them. Not to speak of more indirect pathways: IAA for example is known to be in some way modulated by light. The various transduction pathways are interconnected. Moreover, some intermediary feed back to the PM, ensuring a well controlled transducing system. Thus, the PM is limiting the chances of untimely triggering a particular morphogenetic process: the response is coordinated.

In addition, we propose now that the plasma membrane imposes levels of activity by CR, thus providing a more stringent regulatory control on transduction and ensuing morphogenetic response.

Note Added in Proof. The authors of chap.7 questioned the photoreceptors as single blue light receptor. However our results ascribing CR synchronization and morphogenesis developmental signals to the plasma membrane located photoreceptors in Acetabularia remain probably correct: altough the ambiguity of the salicylhydroxamic acid biochemical effects is not clarified, but it seems that its remarkable disruptive effects on the plasma membrane, together with the circadian time-dependent (but delayed) effects of SHAM on cap formation can be only explained by the disruption of membrane located receptors or associated molecules. It is highly desirable to confirm the interpretation by fluorescence experiments (cf Short and Briggs, cited by Asard and Caubergs, chap.7)

References

André B,Scherer GFE(1989) Stimulation by auxin of phospholipase A in membrane vesicles from an auxin sensitive tissue is mediated by an auxin receptor. Planta 185:209-214

Azard H(1991) Blauw licht perceptie by hogere planten.Studies met betrekking tot ATPase activiteit en redox component in de plasmamembraan. Thesis Universiteit Antwerpen

Berridge MJ(1993) Inositol trisphosphate and calcium signalling. Nature 361: 315-325

Berridge MJ,Irvine RF(1989) Inositol phosphates and cell signalling. Nature 341:197-204

Blatt MR,Thiel G(1993) Hormonal control of ion channel gating. Ann Rev Plant Physiol Plant Molec Biol 44:543-567

Castillo FJ,Cotton G,Kevers C,Greppin H,Gaspar T,Vanden Driessche T(1986) Changes in ascorbic acid content and ascorbate peroxidase activity during the development of Acetabularia mediterranea. Differentiation 33:17-23

Caubergs R,Vanden Driessche T,De Greef J(1984) A light-inducible cytochrome b reduction in the green alga Acetabularia. In: Senger H(ed) Blue light effects in Biological Systems. Springer Verlag Berlin Heidelberg:173-176

Chaignon M,Van Langendonckt A,Vanden Driessche T(1993a) IP3 in Acetabularia.I:Evidence for IP3 in A. mediterranea. Arch Internat Physiol Biochim Biophys 101:3

Chaignon M,VanLangendonckt A,Vanden Driessche T(1993b) IP3 in Acetabularia:II Separation of phosphoinositides. Arch Internat Physiol Biochim Biophys 101:4

Chaignon M.Vanden Driessche T(1993) IP3in Acetabularia IV:Evidence for its physiological role as studied by the effects of lindane. Arch Internat Physiol Biochim Biophys 101:4

Clauss H(1977) Photomorphoses in marine algae. Book of Abstracts Ann Europ Symp Photomorphogenesis Bet Dagan N°16

Comolli J,Taylor W,Hastings JW(1994) An inhibitor of protein phosphorylation stops the circadian oscillator and blocks light-induced phase shifting in Gonyaulax polyedra. J Biol Rhythms 9:13-26

Cooke DT,Burden RS(1990) Lipid modulation of plasma membrane-bound ATPases. Physiol Plant 78:153-159

Cotton G,Vanden Driessche T(1987) Identification of calmodulin in Acetabularia, its distribution and physiological significance. J Cell Biol 87:337-347

Ding JP,Pickard BG(1991) Activation of a Ca^{2+} channel by ABA and resultant inhibition of a K^{+} channel by Ca^{2+} in excised plasmalemma of onion epidermis. Plant Physiol 96 suppl.1:137

Geer van der P,Hunter T,Lindberg RA(1994) Receptor protein- tyrosine kinases and their signal transduction pathways. Ann Rev Cell Biol 10:251-337

Glory M,Vanden Driessche T(1985) NAD^{+}-induced ultrastructural circadian variations in the chloroplasts of Acetabularia.In Bonotto S,Cinelli F,Billiau R (eds) Acetabularia 1984 Belgian Nuclear Center Mol: 267-278

Goldberg RB,Paiva de G,Yadegari R(1994) Plant Embryogenesis: zygote to seed. Science 266:605-614

Goodwin BC and Pateromichelakis S(1979) The role of electric fields, ions and the cortex in the morphogenesis of Acetabularia. Planta 145:427-435

Gradmann D(1976) "Metabolic" action potentials in Acetabularia. J Membrane Biol 29:23-45

Grignon C,Sentenac H(1991) pH and ionic conditions in the apoplast. Ann Rev Plant Physiol Plant Molec Biol 42:103-128

Henry P-E,Bourgeois P,Baeckeland P,Legros F(1979) Relationships between metabolism of carbohydrates and insulin fixation to Acetabularia in LD conditions. In: Bonotto S,Kefeli V, Puiseux-Dao S (eds) Developmental Biology of Acetabularia. Elsevier /North Holland Biomed. Press Amsterdam New York Oxford: 283-294

Horemans N,Asard H,Caubergs R(1993) Ascorbate free radicals possibly act as an electron acceptor to the plant plasmamembrane beta-type cytochrome. Arch Internat Biochim Biophys (in press)

Horemans N,Asard H,Caubergs R (1994) The role of ascorbate free radical as an electron acceptor to cytochrome b-mediated trans-plasma membrane electron transport for higher plants. Plant Physiol 104:1455-1458

Jerebzoff S,Vanden Driessche T(1983) Rythme circadien de la teneur en acides gras dans la cellule entière et les chloroplastes.CR Acad Sc Paris 296: 319-322

Jones AM(1994) Auxin-binding proteins.Ann Rev Plant Physiol Plant Molec Biol 45:393-420

Kevers C,Vanden Driessche T,Gaspar T(1986) An ethylene-forming system in Acetabularia. Relationship with development. Arch Internat Physiol Biochim 94:51

Kevers C,Vanden Driessche T,Gaspar T(1989) Relationship between ethylene and ethane production by whole Acetabularia mediterranea cells and by microsomal fractions. Saussurea 19:121-123

Klämt D(1990) A view about the function of auxin-binding proteins at plasma membrane.Plant Molec Biol 14:1045-1050

Klämt D,Viola G,Barbier-Brigoo H,Guern J(1990) Auxin-binding protein -localization and function. Physiol Plant 79:A32

Kof E,Kefeli V (1979) Growth substances in Acetabularia.In: Bonotto S,Kefeli V,Puiseux-Dao S (eds) Developmental Biology of Acetabularia. Elsevier/North Holland Biomedical Press pp 45-47.

Kosower NS,Kosower EM(1987) Thiol labeling with monobromobimanes. In: Jacoby WB,Griffith OW (eds) Methods in Enzymology Academic Press San Diego. Vol 143:76-84.

Legros F,Saines M,Conard V. (1973) . Action du 3':5'-AMP cyclique et de la théophylline sur la respiration d'Acetabularia mediterranea. Arch Internat Physiol Biochim 81: 164-166.

Legros F,Uytdenhoef P, Dumont I, Hanson B, Jeanmart J,Massant B Conard V(1975). Specific binding of insulin to the unicellular alga Acetabularia mediterranea. Protoplasma 86: 119-134.

Leshem YY(1987) Membrane phospholipid catabolism and calcium activity in the control of plant senescence. Physiol Plant 69: 551-559

Leshem YY(1992) Plant Membranes.Kluwer Academic Publishers Dordrecht Boston London pp 266

Leung J,Bouvier-Durand M,Morris P-C,Guerrier D, Chefdor F, Giraudat J(1994) Arabidopsis ABA response gene ABI1: features of a calcium-modulated protein phosphatase. Science 264: 1448 -1452.

Martiny-Baron G,Scherer GFE(1989) Phospholipid-stimulated protein kinase in plants. J Biol Chem 264:18052-18059

Meyer K,Grill E(1994) A protein phosphatase 2C involved in ABA signal transduction in Arabidopsis thaliana. Science 264, 1452-1455.

Moller IM, Crane FL(1990) Redox processes in the plasma membrane. In:Larson C,Moller IM (eds)The plant plasma membrane.Springer Verlag Berlin Heidelberg New York London Paris Tokyo Hong Kong pp93-126

Mummert H, Gradmann D(1976) Voltage dependent potassium fluxes and the significance of action potentials in Acetabularia.Biochim Biophys Acta 443:443-450

Napier RM,Venis MA(1991) From auxin-binding proteins to plant hormone receptor? Trends Biochemical Science 16:72-75.

Novak B,Bentrup FW(1972) An electrophysiological study of regeneration in Acetabularia mediterranea. Planta 108:227-244.

Novak B,Sironval C(1976) Circadian rhythm of the transcellular current in regenerating enucleated posterior stalk segments of Acetabularia mediterranea. Plant Sci Lett 6:273-283

Ottoline HM,Lincoln CA,Timpte C,Lammer D,Turner J,Estelle M(1993). Arabidopsis auxin-resistance gene AXR1 encodes a protein related to ubiquitin activating enzyme E1. Nature 364:161-164.

Palme K,Schell J(1991) Auxin receptors take shape.Current Biol 228-230

Pause A,Belsham GJ,Gingras A-C,Donzé O,Lin T-A,Lawrence JCjr, Sonenberg N(1994) Insulin-dependent stimulation of a protein synthesis by phosphorylation of a regulator of 5'-cap function. Nature 371:762-767

Penel C,Castille FJ(1991) Peroxidases of plant plasma membranes, apoplastic ascorbate, and relation of redox activities to plant pathology. In: Crane FL,Morré DJ,Löw H (eds) Oxidoreduction at the plasma membrane: relation to growth and transport. CRC Press, Boca Raton. Vol.2:121-147

Puroit S,Kumar P,Laloraya M,Bharti S and Lolaraya MM(1992) Biochim Biophys Res Commun 186:652-658

Reymond P,Short TW,Briggs WR(1992) Blue light activates a specific protein kinase in higher plants. Plant Physiol 100:655-661

Reymond P,Short TW,Briggs WR,Poff KL(1992) Light-induced phosphorylation of a membrane protein plays an early role in signal transduction for phototropism in Arabidopsis thaliana. Proc Natl Acad Sci USA 89:4718-4721

Rubinstein B,Luster DG(1993) Plasma membrane redox activity: components and role in processes. Ann Rev Plant Physiol Plant Molecul Biol 42:107-130

Schaewen von A,Sturm A,O'Neil J,Chrispeels MJ(1993) Isolation of a mutant Arabidopsis plant that lacks N-acetyl glucosaminyl transferase I and is unable to synthetize Golgi-modified complex N-linked glycans. Plant Physiol 102:1109-1118.

Scherer GFE,André B(1989) A rapid response to a plant hormone: auxin stimulates phospholipase A2 in vivo and in vitro. Biochem Biophys Res Commun 163:111-117

Schmid R(1984) Blue light effects on morphogenesis and metabolism in Acetabularia. In:Senger H (ed) Blue light effect in Biological systems. Springer, Heidelberg Berlin:419-432

Schroeder JI, Hedrich R(1989) Involvement of ion channels and active transport in osmoregulation and signaling of higher plant cells. Trends in Biochem Science 14:187-192

Schröder-Lorenz A,Rensing L(1987) Circadian changes in protein-synthesis rate and protein phosphorylation in cell-free extracts of Gonyaulax polyedra. Planta 170:7-13

Schulman H(1991) Serine/threonine kinases in the nervous system. Current opinion in Neurobiology 1:43-52

Shenolikar S(1994) Protein serine/threonine phosphatases-New avenues for cell regulation. Ann Rev Cell Biol 10:55-86

Siegelbaum SA(1994) Ion channel control by tyrosine phosphorylation. Current Biol 4:242-245

Short TW,Reymond P,Briggs WR(1993) A pea plasma membrane protein exhibiting blue light-induced phosphorylation retains photosensitivity following triton solubilization. Plant Physiol 101:647-655

Stemmer P,Klee CB(1991) Serine/threonine phosphatases in the nervous system.Current Opinion in Neurobiology 1:53-64

Sussman MR(1994) Molecular analysis of proteins in the plant plasma membrane. Ann Rev Plant Physiol Plant Molecul Biol.45: 211-234

Theologis A,(1993) What a gas! Current Biology 3: 369-371.

Short TW,Reymond P,Briggs WR(1993) A pea plasma membrane protein exhibiting blue light-induced phosphorylation retains photosensitivity following triton solubilization. Plant Physiol 101:647-655

Vanden Driessche T(1979) Phase-shifting effect of IAA on the photosynthetic circadian rhythm of Acetabularia. In:Bonotto S, Kefeli V,Puiseux-Dao S.Developmental Biology of Acetabularia. Elsevier pp 195-204

Vanden Driessche T(1984) Temporal morphology and cap formation in Acetabularia-Effects of morphactin and auxin. Chronobiol internat 1:113-120.

Vanden Driessche T(1985) Circadian time-dependent effects of plant growth regulators on morphogenesis in Acetabularia. In: Rensing L, Jaeger NI. Temporal Order pp 211-212

Vanden Driessche T(1994) Circadian rhythms in three unicellular organisms. The peculiarities of the organisms, the evidence they brought on rhythms and their specific practical problems. Outline of recent hypotheses. J Rhythm Research 25:345-386

Vanden Driessche T,Caubergs R(1985) Inhibiting the transduction of blue light signals in Acetabularia. In:Bonotto S,Cinelli F, Billiau R (eds). Acetabularia 1984. Centre d' Energie nucléaire

Vanden Driessche T,Delegher V(1977) The presence of an auxin-like substance in Acetabularia and its relation to growth and morphogenesis. In: Woodcock CLF (ed) Progress in Acetabularia Research Acad Press 287-303.

Vanden Driessche T,Chaignon M,Potter BVL(1993) IP3 in Acetabularia III assessment of IP3 physiological role:effects of phosphothioate (IPS3). Arch Internat Physiol Biochim Biophys 101:12

Vanden Driessche T, Cotton G,Loicq C,Caubergs R(1985) Chronobiological approach to the oscillating biochemical pathways identification involved in cap morphogenesis in Acetabularia. Arch Biol (Bruxelles) 96:504-505.

Vanden Driessche T,Kevers C,Collet M,Gaspar T(1988) Acetabularia and ethylene:production in relation with development,circadian rhythms in emission,and response to external application. J Plant Physiol 133:635-639

Vanden Driessche T,Lause P,Guisset J-L,Glory M and Petiau-de Vries G(1992) Some aspects of glycoprotein metabolism in Acetabularia: spatio-temporal activity of beta-galactosidase and physiological effects of tunicamycin. J Interdiscipl Cycle Res 23:237-251

Vanden Driessche T,Petiau-de Vries G,Guisset J-L(submitted) Differentiation: Acetabularia as a model system.

Vanden Driessche T,Verhoustraeten F,Glory M,Guisset J-L(1993). Circadian time-dependent effects exerted by abscisic acid, a phytohormone, on photosynthesis, non-chloroplast polysaccharide content and on cap formation in Acetabularia. In: Sato S,Ishida N,Ishikawa H (eds) V Internat. Colloquium on Endocytobiology and Symbiosis. Tübingen University Press pp 43-46.

Van Langendonckt A,Vanden Driessche T(1992a) Isolation and characterization of different forms of thioredoxins from the green alga Acetabularia mediterranea:identification of an NADP/thioredoxin system in the extrachloroplastic fraction. Arch Biochem Biophys (USA) 292:156-164

Van Langendonckt A, Vanden Driessche T(1992b) Changes in intracellular distribution of thiol groups during the development of Acetabularia mediterranea. J exptl Bot 43: 1643-1650

Van Langendonckt A,Vanden Driessche T(1994) Effects of blue light on thiol/disulfide ratio in relation with growth in Acetabularia. Arch Internat Physiol Biochim Biophys 102:65

Van Langendonckt A,de Solan C,Jerebzoff S,Vanden Driessche T (1991) Non chloroplastic pyridine nucleotides concentrations in Acetabularia as related with the developmental program. Circadian modulation of the concentrations. J interdiscipl Cycle Res 22:223-236

Venis M(1985) Hormone binding sites in plants. Longman (ed), New York London. pp. 191.
Venis M(1987) Hormone receptor sites and the study of plant development. In: Hoad GV,Lenton JR,Jackson MB, and Atkin RK (eds) Hormone action in plant development. Butterworths.London Boston Durban Singapore Sydney Toronto Wellington. pp 53-61
Venis MA,Thomas EW,Barbier-Brigoo H,Ephritikhine G,Guern J(1990) Impermeant auxin analogues have auxin activity. Planta 182:232- 235
Verhoustraeten F(1987) Effets de l'acide abscisique sur le développement de l'algue Acetabularia. Approche de la modulation circadienne régulant certains processus physiologiques et morphogénétiques. Mémoire présenté en vue de l'obtention du grade de lic. Sc.Zool. Université Libre de Bruxelles
Wan Y,Hasenstein KH(1993) Immunological evidence for ABA receptors in corn roots. Plant Physiol 102(1 suppl.):62

Subject Index

—A—

—B—

—C—

—D—

—E—

—F—

—G—

—H—

—I—

—K—

—L—

—M—

—N—

—O—

—P—

—R—

—S—

—T—

—U—

—V—

—W—

—X—

—Y—

—Z—

Springer-Verlag and the Environment

We at Springer-Verlag firmly believe that an international science publisher has a special obligation to the environment, and our corporate policies consistently reflect this conviction.

We also expect our business partners – paper mills, printers, packaging manufacturers, etc. – to commit themselves to using environmentally friendly materials and production processes.

The paper in this book is made from low- or no-chlorine pulp and is acid free, in conformance with international standards for paper permanency.

GPSR Compliance

The European Union's (EU) General Product Safety Regulation (GPSR) is a set of rules that requires consumer products to be safe and our obligations to ensure this.

If you have any concerns about our products, you can contact us on ProductSafety@springernature.com

In case Publisher is established outside the EU, the EU authorized representative is:

Springer Nature Customer Service Center GmbH
Europaplatz 3
69115 Heidelberg, Germany

Batch number: 10704713

Printed by Printforce, the Netherlands